TRANSFORMING AGRICULTURE AND FOODWAYS

Food and Society

Series Editors: **Michael K. Goodman**, University of Reading, UK and **David Goodman**, University of California, Santa Cruz, USA

This series takes an interdisciplinary and integrative approach to food studies. Authors will undertake critical assessments of key topics, controversies and thematic trends across all aspects of food systems and their diverse geographies in order to map out research agendas that bring cutting-edge perspectives to the study of food.

Cumulatively, these appraisals will create an indispensable library for researchers and food studies courses across the social sciences.

Coming soon

Radical Food Geographies
Theory, Action, and Collaboration for Equitable and Sustainable Food Systems Edited by **Colleen Hammelman, Charles Z. Levkoe** and **Kristin Reynolds**

Exploring food citizenship
Networks and Cultures of Collective Food Procurement
By **Cristina Grasseni**

How to Eat Together
Hospitality, Commensality, Conviviality
By **David Bell**

Find out more

bristoluniversitypress.co.uk/food-and-society

TRANSFORMING AGRICULTURE AND FOODWAYS

The Digital–Molecular Convergence

David Goodman

First published in Great Britain in 2024 by

Bristol University Press
University of Bristol
1-9 Old Park Hill
Bristol
BS2 8BB
UK
t: +44 (0)117 374 6645
e: bup-info@bristol.ac.uk

Details of international sales and distribution partners are available at bristoluniversitypress.co.uk

© Bristol University Press 2024

British Library Cataloguing in Publication Data
A catalogue record for this book is available from the British Library

ISBN 978-1-5292-3146-5 hardcover
ISBN 978-1-5292-3150-2 paperback
ISBN 978-1-5292-3147-2 ePub
ISBN 978-1-5292-3148-9 ePdf

The right of David Goodman to be identified as author of this work has been asserted by him in accordance with the Copyright, Designs and Patents Act 1988.

All rights reserved: no part of this publication may be reproduced, stored in a retrieval system, or transmitted in any form or by any means, electronic, mechanical, photocopying, recording, or otherwise without the prior permission of Bristol University Press.

Every reasonable effort has been made to obtain permission to reproduce copyrighted material. If, however, anyone knows of an oversight, please contact the publisher.

The statements and opinions contained within this publication are solely those of the author and not of the University of Bristol or Bristol University Press. The University of Bristol and Bristol University Press disclaim responsibility for any injury to persons or property resulting from any material published in this publication.

Bristol University Press works to counter discrimination on grounds of gender, race, disability, age and sexuality.

Cover design: blu inc, Bristol
Front cover image: alamy/Scharfsinn

For Gail: Partner in Life

Contents

Series Preface		viii
List of Abbreviations		ix
Acknowledgements		xi
1	Introduction: Technological Convergence and Change in Modern Agro-Food Systems	1
2	Precision Agriculture: Big Data Analytics, Farm Support Platforms, and Concentration in the AgTech Space	12
3	Precision Agriculture: Adoption, 'Re-Scripting', Farmer Identity, Path Dependence, and 'Appropriationism 4.0'	21
4	Alternative Proteins: Bio-Mimicry, Structuring the New Protein Industry, 'Promissory Narratives', and 'Substitutionism 4.0'	36
5	Agri-Biotechnology and the Failed Promises of the Seed-Chemical Complex, CRISPR and Gene Editing, and Regulatory Capture	53
6	Between Physical Space and Digital Space: COVID-19, Platform Capitalism, and Changing Patterns of Food Provisioning	71
7	Conclusion: Continuities in Change and Lost Opportunities	88
Postscript		96
Notes		97
References		106
Index		134

Series Preface

Food and Society: New Directions

Series Editors: Michael K. Goodman, University of Reading, UK, and David Goodman, University of California, Santa Cruz, USA.

Food and Society: New Directions takes an interdisciplinary and integrative approach to food studies and the inextricable relationships of food, society and nature. Authors undertake critical assessments of key topics, controversies and thematic trends across all aspects of food systems and their diverse geographies. The series maps out a research agenda that brings cutting-edge perspectives to the study of food through outstanding scholarship presented in an accessible style. The books in the *Food and Society: New Directions* series create an indispensable library for food research and teaching across the social sciences.

List of Abbreviations

AGRA	Alliance for a Green Revolution in Africa
AI	Artificial intelligence
AP	Alternative proteins
APHIS	Animal and Plant Health Inspection Service, US Department of Agriculture
AKIS	Agricultural knowledge and innovation system
Bt	*Bacillus thuringiensis*
CAFO	Concentrated animal feeding operation
CFS	Committee on World Food Security
CGIAR	Consultative Group on International Agriculture Research
COP26	26th United Nations Climate Change Conference
COVID-19	Coronavirus disease
CRISPR	Clustered regularly interspersed short palindromic repeats
CRISPR-Cas9	CRISPR associated protein
DNA	Deoxyribonucleic acid
ECJ	European Court of Justice
EPA	US Environmental Protection Agency
EFSA	European Food Safety Authority
FAO	United Nations Food and Agriculture Organisation
FBS	Foetal bovine serum
FDA	US Food and Drug Administration
GCFI	Gross cash farm income
GE	Genetic engineering
GHG	Greenhouse gas
GM	Genetically modified
GMO	Genetically modified organism
GPS	Global positioning system
HLPE	High Level Panel of Experts on Food Security and Nutrition
IATP	Institute for Agriculture and Trade Policy
IAASTD	International Assessment of Agricultural Knowledge, Science and Technology for Development

ICT	Information and communications technology
IPCC	United Nations Intergovernmental Panel on Climate Change
IPES-Food	International Panel of Experts on Sustainable Food Systems
IPO	Initial Public Offering
LCA	Life cycle assessment
M&A	Mergers and acquisitions
MTA	Material transfer agreement
NBT	New breeding technique
NIH	National Institutes of Health
OECD	Organisation for Economic Co-operation and Development
OPEC	Organization of the Petroleum Exporting Countries
PA	Precision agriculture
RNA	Ribonucleic acid
RTDS™	Rapid Trait Development System
SARS	Severe Acute Respiratory Syndrome
SARS-CoV-2	Severe Acute Respiratory Syndrome coronavirus 2
SCP	Single cell protein
SDGs	United Nations Sustainable Development Goals
SFT	Smart farming technologies
SPI	Science–policy interface
UFCW	United Food and Commercial Workers union
UNCTAD	United Nations Conference on Trade and Development
UNEP	United Nations Environment Programme
UNFSS	United Nations Food Systems Summit
VRT	Variable rate technology
WEF	World Economic Forum
WFP	World Food Programme

Acknowledgements

In writing this book, I am grateful to two close friends and long-time colleagues: John Wilkinson and Mike Goodman. I have benefited from conversations and collaborations over many years with John, and this book is strongly influenced by his outstanding work on the international political economy of agriculture and food systems.

Mike Goodman has been an invaluable guide to the literatures of food studies and geography, unearthing contributions that I would otherwise have missed. His critical readings of preliminary drafts and overall encouragement have been vital in enhancing the quality of this book and bringing it to fruition.

I am also heavily indebted to the 'invisible college' of online journalists for their insightful analysis of developments in the spaces of AgTech and FoodTech. These contributors sadly go largely unrecognised in academic circles, but their work is evident throughout this book.

Finally, I wish to thank three anonymous reviewers for their constructive criticism and suggestions.

1

Introduction: Technological Convergence and Change in Modern Agro-Food Systems

Introduction

This book explores the socio-economic impacts and trajectories of techno-scientific change in the agro-food systems of post-industrial economies in the Global North, notably the United States, the European Union, and the United Kingdom. The aim is to understand how these future agro-food 'worlds' are being socially constructed, their likely evolution, and global implications. Our central theme unfolds against the background of two planetary crises threatening our very existence: global climate change and COVID-19. These crises weave in and out of the analysis with varying degrees of emphasis but are a constant presence throughout. Their damaging effects on food security, nutrition, and human health were compounded in 2022 by the Russia–Ukraine war that has triggered a severe global 'cost of living' crisis, galvanised by inflationary increases in the prices of food and fuel.

An intense wave of innovation is now sweeping over modern food systems,[1] drawing momentum from the dynamic *convergence* of digital code as software-enabled information and communications technologies (ICTs) and the genetic code as molecular engineering technologies. The hybrid techno-scientific and socio-ecological progeny of this convergence are the bearers of our agro-food futures and therefore involve socio-political choices, whether tacit or overt, about the world we are making and are choosing to live in. Technologies, in other words, are irrevocably political.

From this perspective, some contributors to critical food studies argue that the social implications of innovation pathways should be evaluated in anticipatory and more socially inclusive ways within a framework of 'responsible innovation' (Bronson, 2018, 2019; Rose and Chivers, 2018). This proposal is complemented by Carolan's (2018) insistence on a performative approach to agro-food technology that focuses not on what it is but rather on the societal consequences engendered by these emerging socio-technical forms. This critical theme runs through the book but is especially prominent in the analyses of precision agriculture (PA), alternative proteins (AP), and biotechnology.

The convergence of digital and molecular technologies in modern agro-food systems is contextualised by the political economic dominance of oligopolistic agribusiness corporations – Big Ag – whose concentrated power over agriculture R&D reproduces paths of change that are incremental and evolutionary rather than revolutionary in nature. Typically, these trajectories replicate and promote the practices, values, norms, social relations, and power structures of the dominant techno-economic paradigm, here fossil fuel-dependent industrial agriculture. Two concepts from evolutionary economics – *lock-in* and *path dependence* – are used throughout this book to analyse the political economy of the remarkable continuity of industrial agro-food systems.

Powerful modern-day corporations – John Deere, Monsanto, Bayer, AstraZeneca, for example – trace their origins to the birth of industrial agriculture in the later 19th century. This emerging paradigm was cemented definitively in the 1930s and 1940s by an earlier technological convergence: that of mechanical and chemical innovations on hybrid seed in the US Mid-Western Corn Belt (Goodman et al, 1987). This region is now the leading edge of a second transformation as converging digital and molecular innovations redefine production practices and consolidate the incumbent techno-scientific paradigm. These 'paradigmatic technologies' within the dominant socio-technical regime are exemplified by PA and gene editing, as we see in Chapters 3 and 5.

Although we concentrate primarily on 'local' structural problems, these evolving agro-food systems are also fundamental causal factors in the multi-scaled, inter-locking biophysical and socio-economic crises shaping our planetary future. The global climate emergency (Ripple et al, 2020) is the paramount existential question of our time, and the agro-food system is a major source of anthropogenic greenhouse gas (GHG) emissions, led by industrial livestock production with 14.5 per cent of the global total. In these critical circumstances, dietary and systemic policy changes can make a vital contribution to planetary and human health by reducing the consumption of intensively produced ruminants and other livestock (Swinburn et al, 2019).

A Special Report, *Climate Change and Land* (2020) by the UN Intergovernmental Panel on Climate Change (IPCC) estimates that agriculture, forestry, and other land use activities accounted for 23 per cent of total net anthropogenic GHG emissions in the years 2007–2016. A more recent study by the United Nations Food and Agriculture Organisation (FAO) indicates that the global agro-food system emits 31 per cent of total GHG emissions, with pre- and post-production activities playing an increasingly important role since 1990 (FAO, 2021).² The pressure on policymakers and communities to restrict emissions has intensified following the IPCC's *Sixth Assessment Report* (2021), which finds that evidence of human-induced climate change is "unequivocal" and warns of the catastrophic consequences of failure to limit cumulative CO_2 emissions. The UN Secretary General, António Guterres, describes this latest IPCC report as "a code red for humanity" (United Nations, 2021a).

In some perspectives, the global, transnationalised agro-food system is also identified as the systemic cause of successive outbreaks of zoonotic diseases in recent years, such as SARS (Severe Acute Respiratory Syndrome), new strains of swine flu and avian flu, and SARS-CoV-2 (Severe Acute Respiratory Syndrome coronavirus 2), including the strain we know as Coronavirus disease (COVID-19). For example, the Structural One Health approach links the aetiology of disease to ecological destruction, biodiversity loss, and the activities of multinational agribusiness accelerating deforestation and land use change, incorporating new areas into agricultural, often intensive monocultural, production, and destroying natural buffer zones against non-human animal pathogens that can infect humans (Bellamy Foster and Suwandi, 2020; Wallace, 2020, 2021; Lefrancois, 2021; see also Peet and Peet, 2020). In turn, global agro-commodity supply chains and human travel systems act as 'transmission belts' between the deforested tropical areas in the Global South and the leading centres of consumption, finance, and accumulation in the Global North (Bellamy Foster and Suwandi, 2020). "The geography of COVID-19 is produced by globalisation, which links people the world over into a single, unregulated disease system" (Peet and Peet, 2020: 320).

Other analyses of global climate warming and COVID-19 do not take this dialectical, environmental approach but regard the combined epidemiological-economic crises as an episodic event that represents a 'wake-up call' for transformative change of the global agro-food system, whose structural vulnerabilities and frailties have been so starkly exposed (Clapp and Moseley, 2020; Hendrickson, 2020; Herren, 2020; IPES-Food, 2020). At the other end of the spectrum, observers view COVID-19 as a 'stress test' and are impressed by the resilience of transnational commodity chains and distribution systems in withstanding the potentially disruptive impacts of the pandemic (Holmes, 2021).³

The central role of the global agro-food system in these planetary crises, and their cascading consequences on multiple scales, emerges clearly in an earlier IPCC report calling urgently for a global sustainability transition to limit global warming to 1.5 °C above pre-industrial levels (IPCC, 2018). The enormity of this task is indicated by estimates that, if food system-related GHG emissions were to continue at their current level, these alone would consume the *entire* allowable emissions budget consistent with achieving the targets set out in the 2015 Paris Agreement (Clark et al, 2020; see also Moran, 2021) Some observers expect animal-free, alternative proteins, the 'poster child' of GHG mitigation for Silicon Valley hedge funds, to play an important role in reducing the size of the livestock industry. A variety of views on this issue, including the prospect of a privatised alternative protein supply chain enmeshed in intellectual property rights, are discussed in Chapter 4.

This focus on industrial innovation to mitigate global warming stands in contrast to the rising emphasis on agro-ecological pathways to sustainable agro-food systems in the narratives of some multi-lateral agencies and civil society actors. However, there are few concrete signs that modern agro-food systems are in the vanguard of a sustainability transition to a lower carbon future. On the contrary, the 'two code convergence' is reinforcing the hegemonic socio-technical regime of fossil fuel-based industrial agriculture, aka 'productivist' agriculture, and the market power of its foundational 'heritage' companies. Of course, Western agro-food systems are diverse, with active sectors of organic, agro-ecological, local, and 'regenerative' agriculture, and these may all form part of an eventual transition. Nevertheless, for the present, mainstream commodity agriculture is dominant and remains 'locked into' ecologically damaging, socially dis-equalising practices. Despite the rhetorical flourish of such labels as 'climate smart' PA or 'sustainable intensification' and reference to 'disruptive', paradigm-changing technologies, there currently is little convincing evidence that industrial commodity agriculture is embarked on a sustainability transition, even in pre-figurative form.

The continuity and consolidation of the oligopolistic corporate structures that have long dominated the industrial food regime exemplify the path dependent nature of current agricultural techno-scientific innovation. Indeed, using a variety of market entry strategies, established corporate actors have surfed the wave of innovation to retain control of agriculture R&D, exploit new sectors of value creation, and articulate new identities. Oligopolistic meat processors, such as JBS Foods and Tyson Foods, for example, now marketing plant-based meat substitutes, have become self-styled 'protein companies', while life science corporations are represented as 'information companies' following their diversification into digital farm management. Belying their size, long-established hegemonic actors have shown agility in diversifying into the alternative proteins industry and adopting the digital platform business model to ensure that they are in the ranks of the "'disruptors' rather than the 'disrupted'" (Stephens et al, 2019: 7).

Continuity also characterises the deep-seated structural fault lines and ecological problems of industrial agriculture in the Global North, as discussed in Chapters 2 and 3. This historic legacy is invoked vividly by the Institute for Agriculture and Trade Policy (IATP, 2020), when it revisited a foundational analysis of the 1980s US farm crisis. The fault lines exposed in 1987 by this earlier IATP diagnosis called 'Crisis by Design' – farmer indebtedness, bankruptcies, farmland consolidation, the loss of rural livelihoods, corporate concentration, moribund rural communities, race and gender inequality, social injustice, and environmental degradation – have all persisted and intensified. The failure to resolve these rising tensions within industrial food systems, now heightened by the greater frequency and intensity of extreme climate events, provides the 'local' context of technological innovation presented in this book.

Corporate interests seek to dissipate these tensions, with discourses supporting technocratic, market-based solutions and re-asserting the benefits of greater capitalisation, effectively endorsing the 'business-as-usual' approach that created the fault lines in the first place. A common stratagem is to extol the role of the technologically 'efficient' farmer in averting the danger of global hunger by maximising production to 'feed the 9 billion' by 2050, a discourse described by Weis (2015) as the "doubling narrative". In this way, contentious politics are accommodated and diffused within existing power structures and translated into manageable technocratic responses (see also Newell, 2017).

Concepts and themes

A previous book on innovation and agro-industrial development, *From Farming to Biotechnology* (1987), written with Bernardo Sorj and John Wilkinson, was built around the centrality of space and time in the capitalist commodification of the farm labour process, which we conceptualised as Appropriationism. This analysis was strongly informed by the work of economic historians, such as Allan Bogue, with his eloquent imagery of the industrial appropriation *via* mechanisation of the small grain harvest in the US Mid-West:

> The man who could sickle less than an acre of grain in a day could cradle [scythe] at least two. place him on the seat of a Virginia reaper, and he could cut 10 to 15 acres per day, with the aid of one man to rake the grain clear of the table. By the 1860s the self-rake reaper had eliminated this helper from many harvest fields. (Bogue, 1968: 164)

Following historically discontinuous but progressive steps, elements of the production process – traction, soil nutrients, seeds, for example – were appropriated by industrial capitals and re-integrated as *externally produced farm inputs*.

As Leszczynski (2019: 13) observes, "Geographers have (long) recognised technology to be central to the production of space and socio-spatial relations [spatiality]". Or, more specifically:

> Space is not simply a container in which things happen; rather spaces are subtly evolving layers of context and practice that fold together people and things and actively shape social relations
> [...]
> society, space and time are co-constitutive – [that is,] processes that are at once social, spatial and temporal in nature and produce diverse spatialities. Software and the work it does are the products of people and things in time and space, and it has consequences for people and things in time and space. (Kitchin and Dodge, 2011: 13)

In analysing the structural trajectories of modern food systems, we draw on these and other recent contributions in digital geography that theorise the interactions between technology, space, time, and socio-spatial relations as co-productions (see Ash et al, 2019).

These co-productions describe the 'Appropriationist' pathways linking socio-spatial assemblages in the historical progression from hand tools and horse-drawn machinery to industrial modernisation with tractors, combines, hybrid seeds, and agro-chemical practices, the energy transition from biomass to fossil fuels, and contemporary technologies of digitalisation and genetic engineering (GE). These pathways are also marked by the structural fault lines identified by the IATP (2020) associated with the continuing treadmill-like commodification and capitalisation of the farm production process and the progressive fall in worker-hours per acre this entails (Bogue, 1968, 1983; Cochrane, 1979).

Now, rather than the mechanisation of extensive space, 'Appropriationism 4.0' works in digital space by 'miniaturising' farm fields into high-resolution, grid-like patterns using global positioning system (GPS)-assisted technologies to collect 'Big Data' for analytic processing, in order to determine and optimise site-specific requirements of seed, fertilizer, crop protection chemicals, and related inputs. In effect, farmers' tacit, experiential knowledge of farm space is displaced and commodified by corporate digital support platforms and integrated into the production process as algorithmically prescribed input applications. The changing materiality, spatiality and 'ways of knowing' the farm and the corresponding 'relocation of power' (Gibson, 2019) are discussed in detail in Chapters 2 and 3.

In the previous book, Appropriationism was complemented by the concept of Substitutionism to capture the reductionist processes transforming traditional rural whole foods and crops into interchangeable inputs or 'intermediates' for use as ingredients in the manufacture of industrial foods.

Following the introduction of mechanised food processing between 1870 and 1914, "the food industry subsequently could turn its attention to effecting qualitative changes in the organic composition of food and the general perception of what constitutes food" (Goodman et al, 1987: 60). This earlier analysis traced the deployment of separation and fractionation methods and industrial biotechnologies in the production of highly processed, fabricated foods endowed with industrial attributes, such as shelf-life, convenience, specific nutritional qualities, taste, texture, and flavours. The latest iterations of these molecular techniques underpin the alternative proteins industry discussed in Chapter 4.

The narrative structure of this book does not present a singular conceptual framework, nor follow a central unifying thread linking its chapters. Rather, to understand the complex, co-produced assemblages forged by the integration of the digital and biotechnological, its theoretical and empirical fabric is woven from multiple sources. It builds from the foundations of Appropriationism and Substitutionism to draw conceptual strands from science and technology studies, particularly actor-network theory, neo-Schumpeterian evolutionary economics, and analyses of inter-capitalist competition. For example, in the analysis of farmers who are interrogating the 'prescriptions' of digital farm service platforms, we follow David Rose and his colleagues in their use of the Latourian concept of 're-scripting'.

Evolutionary economics provides a conceptual toolkit to understand the mechanisms that determine the persistence and reproduction of dominant socio-technical regimes (Dosi, 1982; Vanloqueren and Barat, 2009; Bruce and Spinardi, 2018). In particular, this theoretical approach recognises how a constellation of technological, cognitive, epistemological, and institutional factors can combine so that a certain technological approach becomes locked-in, excluding alternatives, creating a path-dependent technological trajectory of incremental rather than radical paradigmatic change. The concepts of lock-in and path-dependence are vital to an analysis of the continuity of the industrial food system and its corporate architecture, as already noted.

Our use of the notion of inter-capitalist competition is more generic and focuses largely on the strategies marshalled by large corporate actors to advance and maintain high levels of industrial concentration. These strategies range from constructing barriers to entry based on intellectual property rights, as in the case of the life science corporations, to a variety of instruments used to gain access to new sectors, keep abreast of technological change and avoid losing market share to smaller innovative competitors. These include mergers and acquisitions (M&As) to takeover or acquire an equity stake in rival firms, participation in the early funding rounds of innovative start-up firms, and in so-called 'incubator' and 'accelerator' funds.

Many large corporations have a dedicated venture capital arm for precisely this purpose.

While these instruments are widely utilised in digital sectors, platform markets have different competitive dynamics that promote 'winner-take-all' strategic behaviour based on their singular structural characteristics, which we identify in Chapter 2. This race for sector dominance is seen in the analysis of corporate farm service platforms, and downstream in the accelerated rise, driven by the COVID-19 pandemic, of digital home food delivery platforms and on-demand food services generally, as discussed in Chapter 6.

Chapter previews

2 Precision Agriculture: Big Data Analytics, Farm Support Platforms, and Concentration in the AgTech Space

The historical re-ordering of space-time and socio-structural transformation of the small grains harvest in the Global North since the mid-19th century can be seen as a succession of inflection points, with digitalisation of major field crop production now the latest of these, actively forging new socio-spatial structures in the same way as its predecessors. PA, aka 'smart farming', involves the diffusion of GPS-assisted mapping tools, GPS auto-steer farm equipment, and site-specific variable rate technologies for input applications. These innovations are instrumental in the transition from field-level management to high-resolution 'grids' at a sub-field scale.

This chapter discusses the role in this transition of Big Data analytics and digital farm management platforms, whose growth set the stage for a series of recent mega-mergers in the 'seed-chemical complex' of agricultural life science companies. This round of consolidation was triggered by Monsanto's acquisition of Climate Corporation in 2013 and this complex is now dominated by the 'Big Four'.

A key argument is that PA technologies and corporate management support platforms offering algorithmic 'prescriptions' for variable rate input applications have reinforced farmer lock-in to the hegemonic paradigm of 'productivist' agriculture.

3 Precision Agriculture: Adoption, 'Re-scripting', Farmer Identity, Path Dependence, and 'Appropriationism 4.0'

This chapter examines the specificities of the diffusion processes of PA technologies and the contested issues these reveal. These specificities are seen in the context of structural agrarian change over the past 30 years, with its distinctive 'divides' and inequalities.

Digitalisation and decision support tools are imbuing farms with certain features of digital control centres by changing the spatial nature of farm work as screen time becomes increasingly important in managing farmscapes. This chapter explores this re-definition of space, time, and socio-spatial relations, and the implications for farmer autonomy and the politics of knowledge production, including the question of path-dependence, and discusses the 'right to repair' controversy as an expression of farmer disempowerment.

PA technologies are transforming farmers' experiential knowledge into newly commoditised inputs, transferring power to corporate service platforms and prescriptive software packages. Farmers' equivocal engagement with these technologies is analysed using the concept of 're-scripting'. We conclude with some reflections on 'Appropriationism 4.0'.

4 Alternative Proteins: Bio-Mimicry, Structuring the New Protein Industry, 'Promissory Narratives', and 'Substitutionism 4.0'

This chapter analyses the rise of the alternative proteins industry, which provides a second vivid example of changing spatialities in modern food systems. Plant- and cell-based protein analogues threaten to displace the farmed livestock industry by compressing time and space, redefining the meaning of urban and rural. Later sections analyse the responses of Big Meat and Big Food corporations to these innovations and the formation of 'protein conglomerates'.

AP start-ups have attracted massive funding from venture capitalists using 'promissory narratives' to claim that meat substitutes can contribute significantly to the reduction of anthropogenic GHG emissions and address global hunger. These discursive claims are interrogated in the light of the recent disappointing performance of the plant-based sector, estimates of the infrastructural requirements needed to make even modest inroads into the global protein market, and recent critiques of the simplistic argument that prioritises a protein transition over a systemic sustainability transition.

The chapter concludes with reflections on how the AP industry replicates long-standing trajectories in food science and manufacture to redefine the terms of industrial dependence on agricultural materials, a process conceptualised as 'Substitutionism'. A final discussion emphasises that this industry is intensely proprietary, with the model of food-as-software secured by food-as-intellectual property.

5 Agri-Biotechnology and the Failed Promises of the Seed-Chemical Complex, CRISPR and Gene Editing, and Regulatory Capture

This chapter traces the consolidation of power over the global seed-chemical complex by an elite group of 'legacy' life science corporations and the parallel privatisation of agricultural R&D. These developments have reinforced the industrial agro-foodsystem, reproducing its path-dependent techno-scientific trajectories and exacerbating the structural crises of rural economy and society. The 'Big Four' life science corporations have erected formidable barriers to entry using intellectual property rights platforms, which dominate the innovation processes in agricultural biotechnology.

This political economy is explored over several generations of plant biotechnologies, beginning with their commercial introduction in the 1990s, the emergence of gene editing systems, notably CRISPR associated protein (CRISPR-Cas9) after 2010, and concluding with the advent of gene driving. Gene drives radically extend the scope of molecular intervention from individual field crops and animal species to 'engineering' agro-ecosystems, and potentially to global scales.

This chapter also analyses the contrasting trajectories of the regulation and governance of biotechnology in the US and the EU, and argues that the US is an exemplary case of regulatory capture by industry.

6 Between Physical Space and Digital Space: COVID-19, Platform Capitalism, and Changing Patterns of Food Provisioning

This chapter focuses on the downstream food system and examines how digitally mediated, platform-based technologies are reconfiguring food services, retailing, shopping practices, and foodways, and how certain earlier trends have been magnified by COVID-19. Big-box grocery stores have adapted relatively well to the rise of online, contactless shopping in contrast to the bleak experience of dine-in restaurants, which suffered from the extraordinary expansion of app-based home delivery platforms in 2020–2021.

This discussion explores the 'winner-take-all' dynamic of platform capitalism in the context of the vertical integration and global consolidation of these delivery platforms. The analysis draws attention to the flawed business model of these sectors, and takes us into the world of 'dark' stores, 'ghost' kitchens, and on-demand convenience, which rely on precarious 'gig' economy employment practices of casualisation, piece rates, and limited access to labour benefits. Recent attempts to re-classify these workers as employees rather than independent 'contractors' are reviewed, notably California's Proposition 22 and the European Commission's decision to issue draft regulations supporting re-classification.

Finally, this chapter examines the COVID-19 pandemic through the lens of global socio-economic breakdown, exposing the frailty of public safety nets, long-standing social inequalities in food security and access to nutritious food, and the racial and ethnic injustices embedded in food and economic systems alike. Looking forward, the dramatic changes in consumer practices in 2020–2021 are becoming embedded in social foodways and, with the prospect of new COVID-19 variants, this structural transformation of food provisioning and eating habits is likely to be consolidated in the years to come.

7 Conclusion: Continuities in Change and Lost Opportunities

This chapter recapitulates the book's main arguments and empirical findings, highlighting the striking continuity of the unsustainable hegemonic regime of fossil fuel-based industrial agriculture in an era of global heating and existential crises for humanity

The innovations analysed in this book, including PA, alternative proteins, gene editing, and home food delivery platforms, are easily accommodated in this hegemonic system. In short, the wave of innovation represents evolutionary rather than radical change. The counterpart of this central continuity is the lack of progress towards a broadly-based sustainability transition.

This impasse is reflected in the deeply entrenched power structures that thwarted hopes of reaching a progressive compromise at the 2021 United Nations Food Systems Summit. This summit was an uninspiring prelude to the United Nations Climate Change Conference (COP26), which generated a range of nonbinding commitments but few concrete measures to mitigate climate change.

A postscript briefly discusses the repercussions of the renewed Russia–Ukraine war on food security and health as food and fuel price inflation accelerate on a global scale, threatening international economic recession.

As these chapter previews suggest, this book explores the co-produced assemblages of social relations, practices, space, time, and nature engendered by two converging codes: the digital and the genetic. This is just a way of saying that modern industrial agriculture is becoming a dual coded space, mutually constituted by software-enabled technologies and gene edited crops. These assemblages also reveal the embryonic form of possible future agro-industrial worlds: fully automated farm production, and mono-cultural agro-ecosystems 'sanitised' by gene drives.

An emphasis on co-production and co-evolution disarms the narrative of the technological 'fix' and the notion that innovation is externally imposed, uncontested, and 'disrupts' a static socio-economic system. Agro-food systems are dynamic, of course, but technological and organisational change typically are gradual and incremental rather than unforeseen and 'revolutionary', as the equivocal embrace of PA clearly demonstrates. As this book reveals, the discourse of disruption normalises the current and future techno-scientific order while conveniently obscuring the continuity of its power structures and embedded injustices.

2

Precision Agriculture: Big Data Analytics, Farm Support Platforms, and Concentration in the AgTech Space

Introduction

In the online reporting and boosting of technological innovation in agriculture, the 'AgTech space' is a broad, amorphous category, but its central frame of reference is the digitalisation of agricultural commodity production. Leading dramatis personae include Silicon Valley venture capitalists financing innovation 'accelerator funds' and 'incubators' that are propelling the explosive growth of start-ups, mainly software firms – Day (2019) has mapped over 1,600 start-up companies in this sector in the past 5 years. This innovation space also is populated by Big Ag and Big Tech mega-platforms[1] pursuing M&A strategies and joint ventures, discussed later.

In this and the following chapter, we examine the varied dimensions and implications of the digitalisation of major field crop production in the US and Western Europe. Cumulatively, these analyses support our central argument that current techno-scientific innovation is path dependent and reinforces the incumbent paradigm of industrial agriculture, as Wolf and Buttel (1996) concluded much earlier. Moreover, the diffusion of PA technologies and the accompanying emergence of corporate farm support platforms have provoked new rounds of industrial consolidation, accentuating already high levels of concentration in the 'seed-chemical complex' (Howard, 2016, 2021; MacDonald, 2019). These points, together with a brief digression on platforms as an organisational form and business strategy, are developed more comprehensively in the following sections.

The key digital innovations of PA are GPS-assisted systems for yield monitoring and soil mapping, auto-steer farm machinery guidance systems, a variety of sensors, and surveillance devices, such as drones, that combine to support site-specific, variable rate input application technologies. In practice, these technologies are being adopted and disseminated, whether singly or in combination, at different speeds, depending on farm size and farmer risk preference. Access to such so-called decision support tools is controlled by the proprietary farm management platforms of competing agribusiness corporations, including Bayer-Monsanto, the Syngenta Group, and John Deere. The adoption and diffusion process and contested issues of farmer identity, privacy, data ownership, and technological path dependence are taken up in Chapter 3.

These technologies are now synonymous with PA, aka 'digital farming' and 'smart farming'. That is, a system of farm management that uses powerful digital and geo-spatial technologies to collect and analyse high-resolution data to guide decision-making at the field and sub-field level. GPS-assisted analytical tools can delineate management 'zones' or 'grids' based on high-resolution soil and yield maps to prescribe variable rate applications of seeds, plant protection chemicals, and fertilizers.

As we will see, the co-production known as precision agriculture and the values and social realities it propagates is radically re-ordering the socio-spatial relations of commodity agriculture. On-farm work practices are changing as knowledge production is increasingly *devolved* to decision support platforms, defining a further round of commodification and the treadmill of competitive innovation.

Digitalisation, Big Data, and digital support platforms

Precision agriculture emerged in the mid-1980s with the development of a rapidly diversifying suite of remote sensing techniques, including GPS-satellites, multispectral aerial drones, and tractor-mounted sensors. John Deere installed GPS software for autonomous, self-steering tractors in the early 1990s and the first crop monitoring sensors followed in 1995.[2] This re-structuring of the farm production process and its spatiality (Rose et al, 2018) has continued to gather pace, drawing particular impetus from Big Data analytics, discussed more specifically later. Sophisticated but relatively cheap hardware sensors and an array of software-enabled devices have automated the collection across farm landscapes of large-scale datasets. These are uploaded to cloud-based storage and computing centres, such as Amazon Web Services, and processed in nearly real time using predictive and prescriptive Big Data analytic software, AI, and machine-learning algorithms.[3]

For Sonka (2016: 7), Big Data is not an entity but a dynamic, expanding capability derived from several technologies that "include, but are not limited to, computation, data storage, communications, and sensing". This analytic capability often is described by the traits of volume, velocity, and variety – the three Vs – applied to large-scale datasets whose dimensions place them "beyond the ability of typical data software tools to capture, store, manage and analyse" (8).[4] But it is analytics – the "hidden 'secret sauce'" (Sonka, 2016: 8) – that truly sets Big Data apart: the power to extract real time descriptive, diagnostic, predictive, and prescriptive findings of commercial value.

Supported by the affordances of Big Data analytics, PA configures a new corporate arena or "economic space" (Kenney, 1998), where agro-industrial capitals compete to become the dominant digital platform for the delivery of decision support tools and sale of the prescribed physical inputs. These services are marketed on proprietary platforms, the digital interface between agribusiness and client-farmers, who can access *personalised* operational advice to inform their farm management decisions.[5] As one farmer who tested the John Deere Green Star Yield Monitor and Mapping System stated approvingly, "When we brought the card in from the combine, we could get a map at the end of the day to see how well our fields yielded" (cited in Stine, 2019).

At the risk of repetition, the technologies of PA promise to achieve cost- and land use-efficiencies by determining location-specific, variable rate applications of agro-industrial inputs. In the words of an Illinois farmer, "The aim is to *customise* every input to every field and each zone within the field, and to have enough data and analytic tools and models to make the best farming decisions possible" (cited by Plume, 2014b; my emphasis). The mainstream diffusion of PA practices also has benefitted from complementary advances in more broadly-based technologies, including high-speed internet, smart phones, apps, low cost and reliable satellites, and improved, variable rate technology (VRT)-enabled farm equipment (Schmatz, 2017).

For these reasons, the 'intelligent tractor' is seen as the 'gatekeeper' of agricultural digitalisation, giving farm equipment firms a strong competitive edge over agro-chemical and life science companies in the struggle to restructure the input industries. As the ETC Group (2016a) notes, "They have the 'box' that applies the seeds, pesticides, fertilizers and water in the field at the beginning of the season and harvests the crop at the end of the season." Furthermore, the sales revenues of farm equipment companies dwarf those of life science corporations, with John Deere's farm machinery sales almost equalling the combined sales of the (then) Big Six seed and agro-chemical companies (IPES-Food, 2017).

However, the outcome of this struggle is by no means a foregone conclusion, as agro-chemical and life science firms quickly recognised that digital service platforms play a key role in product sales. As Steve Cubbage of the farm data management firm, Farmobile, observes:

Companies are basically in a race to gather as many total acres of data, ingest it into the system, and I don't know if anybody knows exactly what they are going to do with the data. They know if they've got it, they're in control. It's a giant land grab. (Cited by Bloch, 2019)

Platforms and digital markets: a brief digression

In view of the rising importance of digital platforms throughout modern food systems, it is worth pausing here to recognise their *sui generis* nature as an organisational form and to emphasise that digital markets are characterised by modes of competition and dynamics of concentration unlike those found in the analogue world. For Coyle (2018: 51), platforms are 'hybrid entities', since "a platform is a business strategy as much as a kind of organization, and some firms operate both one- and two-sided lines of business (such as Amazon as a retailer and Amazon Marketplace)." It is these market specificities and the digital revolution that have drawn such close attention to the global prominence of Big Tech – Google, Apple, Facebook, Amazon, Microsoft and their Chinese counterparts, such as Alibaba and Tencent (see also Moore and Tambini, 2018; Cusumano et al, 2019). Using Big Data technologies to extract value from the collection and analysis of large, complex data sets, these companies have created new 'market spaces' adapted to different ways of doing business and changing the strategic content of market power.

Analysing the market dominance achieved by Big Tech, Barwise and Watkins (2018: 28) emphasise that Big Data analytics techniques "offer both revenue and cost economies of scale, scope and learning". These scale economies are "extreme" and a central reason why "winner-take-all" is the characteristic mode of dominance in digital markets (Srnicek, 2017a; Moore and Tambini, 2018; Cusumano et al, 2019). Analysing this intrinsic feature of these markets, Barwise and Watkins (2018, 22) identify a long-term pattern in new technology sectors "which typically start highly contested but soon become dominated by one (or two) US companies". For Srnicek (2017a: 89), "the platform ... has data extraction built into its DNA" and this imperative defines its competitive logic and dynamics of growth, where "more users begat more users" (45). Such 'network effects' and positive feedback loops confer the potential for explosive, geometric rates of expansion. These structural dimensions of digital markets explain the strong tendency for "early advantages to become consolidated as permanent positions of industry leadership" and power (Srnicek, 2017a: 89).

This digression helps to contextualise the earlier comment by Steve Cubbage of Farmobile on the drive to recruit farmers and their 'acres of data' into farm services platforms: "They know if they've got it, they're in control. It's a giant land grab". In short, as Sykuta (2016: 58) puts it, data service companies are positioning themselves to "harness and commercialise the revenue-generating correlations to be discovered in these fields of data". Realisation of this potential is the driving force behind the reaction of rivals to Monsanto's takeover of the Climate Corporation, as we see later.

With the experience of the Big Tech corporations and the 'winner-take-all' dynamic in mind, farm service platforms use various stratagems to raise switching costs to discourage clients from moving to competitors. As in other digital sectors, 'churn' is to be avoided if at all possible by personalising the customer–provider relationship to lock in users, including offers of products and services at below cost or free of charge. For example, farmers who bought US$45,000 worth of products from Syngenta reportedly received free access to its Agriedge software, a clear attempt to create positive feedback effects to expand its platform (Bloch, 2019). Digital platform markets also are characterised by efforts to 'sign up' clients by 'bundling' a range of services together as a step towards building a 'full stack' technology and AI platform. These pressures are driving diversification strategies, typically through M&A activities and cross-sector alliances, promoting vertical integration, and accentuating concentration in this strategic market.

The farm equipment industry has also adopted platform 'bundling' strategies, as exemplified by John Deere's early response to Monsanto's bid to lead the digital agriculture platform sector by acquiring Precision Planting in 2012, followed by the Climate Corporation in 2013. John Deere had attempted to buy Precision Planting, whose speciality is variable rate input technology, from Monsanto's Climate View subsidiary in 2015, but this move was blocked by the US Department of Justice on anti-trust grounds that one firm would dominate the market for high-speed planting systems. Furthermore, "This buyout would also have allowed Deere equipment to have exclusive, near real-time access to Monsanto's Climate FieldView data (now owned by Bayer), which was used on more than a third of US farmland at the time" (Howard, 2021: 163).[6]

This blocked acquisition also revealed the close relationships between supposed rivals in digital agriculture platforms since customers of Climate Corporation continued to use John Deere's Wireless Data Server technology to stream real-time data from field equipment. According to a Climate Corporation news release in 2017, more than 70 per cent of FieldView drive data were streamed from John Deere planters and combines. Precision Planting was acquired in 2017 by AGCO, one of the powerful triumvirate of leading farm equipment companies.[7]

In the same year, John Deere succeeded in its quest to diversify its PA technologies and extend its platform services by purchasing Blue River Technologies. This Silicon Valley agricultural robotics start-up specialises in computer vision and machine-learning that can be used, for example, in weed detection and herbicide treatment. The next section looks in greater detail at the diversification and vertical integration strategies adopted by agro-chemical and life science companies competing in the market to supply digital support tools.

Precision farming platforms and consolidation in the agricultural life sciences

As occurred in the rise of the 'seed-chemical complex' (Howard, 2015), now formed by the 'Big Four', M&A strategies are at the very heart of their diversification into PA service platforms. This move marks the third phase of a process of vertical integration and industrial concentration that began in the 1980s with the purchase of agricultural biotechnology start-ups and seed companies. A second phase followed in the late 1990s and 2000s with the development of genomics, which cemented the integration between the digital and molecular sciences.[8]

The catalyst for this latest round of consolidation was Monsanto's determined bid to differentiate itself from its rivals by purchasing Precision Planting in 2012 and Climate Corporation for US$1.1 billion in 2013. Strategically, the takeover of Climate Corporation united this company's climate database with Monsanto's unrivalled crop genomic databases (Carolan, 2017b). This 'quantum leap' into the world of PA platforms recognised that success would bring the indirect network effects of greater sales of proprietary seeds and crop protection chemicals. At that time, Climate Corporation, a Silicon Valley start-up and leader in data analytics, owned an extensive database assembled from government weather satellites, weather stations and sensors, and a mobile app, Climate Basic, which provided high-resolution soil and moisture data and other yield indicators. Monsanto later re-launched this app as a more diversified agronomic platform, Climate FieldView, with prescriptive software tools to advise farmers on variable rate seeding and how to match input use to plant needs, as well as data connectivity via Bluetooth and iPad ports in field equipment.[9]

Subsequent company takeovers both by Monsanto and Climate Corporation emphasise the importance of 'bundling' a variety of complementary services to enhance the attractions of these platforms.[10] The acquisition of the Climate Corporation suggested that Monsanto was becoming an "information company", a view reiterated following its acquisition by Bayer AG for US$63 billion in 2018: "The seed giant has spoken of a future where 'the information itself becomes the business'" (cited in Bloch, 2019).[11] 'Bundling', or vertical integration, similarly motivated this mega-merger by combining Bayer's expertise and leadership in crop protection with Monsanto's patented GM seed varieties to provide "a complete crop solution", according to a spokesperson.

This succinct description neatly encapsulates the asymmetric social relations of PA, technological path-dependence, treadmill economics, and the links between digital support platforms and corporate concentration.

The striking takeover of Climate Corporation in 2013 and recognition of the powerful lock-in effects of digital support tools accelerated efforts by rival agro-chemical and life science corporations to emulate Monsanto by acquiring companies specialising in digital agriculture services and software expertise in platform building. Indeed, Leclerc and Tilney (2014) argue that Monsanto's M&A strategy stimulated a momentum shift in the AgTech space, describing 2014 as "a breakout year – the 'Netscape' moment for agricultural technology" – as investment in the sector reached US$2.35 billion spread over 264 deals.[12]

Dan Burdett, head of digital agriculture at Syngenta, put its 2018 purchase of FarmShots, Inc., a company specialising in high-resolution satellite imagery, in the same context as follows: "It's a race to build digital capability. *It's a race to have a digital relationship with the grower.* Whether it's the price paid for Climate (Corporation) or ... for [software company] Granular [by Dow/DuPont], these things are getting everybody's attention" (cited by Cosgrove, 2018). Syngenta, then one of the 'Big Six' agro-chemical and life science companies, bundled FarmShots into its farm management platform, AgriEdge Excelsior, thus offering remote sensing capabilities for the first time, a major step since *"a significant part of our sales go through this platform"* (cited by Cosgrove, 2018; my emphasis). The takeover of FarmShots was complemented by the acquisition of other farm service software providers, including the Brazilian firm, Strider, and AgConnections, while its venture capital arm, Syngenta Ventures, "has backed dozens of start-ups" (Ellis, 2020). In this corporate merry-go-round, Bayer sold its digital assets to BASF as a condition of approval by the US Department of Justice for its 2018 takeover of Monsanto.

Philip Howard (2015) has analysed the development of an integrated 'seed-chemical complex', with M&A activity paring down the plant science and plant protection sectors from 30 firms in the 1970s to the 'Big Six' in 2001. Since then, as we have seen, the rise of closed proprietary platforms as the new business model has intensified pressures towards further concentration in these industries (IPES-Food, 2017). These pressures culminated in the round of 'mega-mergers' in 2016–2018 as the 'Big Six' firms of the 'seed-chemical complex' became the 'Big Four': Bayer/Monsanto, Corteva Agriscience (the spin-off from the Dow/DuPont merger), ChemChina/Syngenta, and BASF. The ChemChina/Syngenta organisation has operated since 2020 as the Syngenta Group, with interests in crop protection, seeds, biotechnology, and fertilizers.[13] These oligopolistic firms account for a combined 60 per cent of global seed sales and over 70 per cent of global pesticide sales (IPES-Food, 2017; Howard, 2018; MacDonald, 2019).

PA technologies and Big Data analytics clearly are disturbing the balance of power in the input industries, and, some would argue, tilting it towards the farm equipment sectors (ETC Group, 2018a; Wilkinson, 2019). The ETC Group (2018a) sees the struggle to control farm data as pitting the 'hardware hub' around farm machinery against the 'software hub' of the 'Big Four' genomics, seeds, and crop protection corporations, with the odds favouring the former. However, the recent mega-mergers in the 'seed-chemical complex', cross-sector alliances between these corporations and John Deere and other farm equipment companies (Bloch, 2019; Howard, 2021), as well as continuing M&A activity to enhance both vertical and horizontal integration and the ability to 'bundle' platform services, make it difficult to be unequivocal about the eventual outcome at this stage.[14] Indeed, as the ETC Group (2018a: 16) acknowledges, "All of the actors at the input end of the industrial food chain, from seeds to fertilizers to machines, are developing Big Data sensors and working with robotics".

The very nature of this debate is quite revealing in itself. A new techno-scientific infrastructure based on digitalised farm machinery and proprietary service platforms clearly is emerging in advanced commodity agriculture. Some observers seemingly take perverse delight in characterising these recent innovations as 'disruptive', as if they are harbingers of paradigm change rather than further evidence of the path dependent continuity of industrial agriculture. This argument is reinforced by the analyses of changes in farm work routines, knowledge production, land consolidation, and the health of rural communities, as we see in Chapter 3.

This comment also obscures a more obvious and significant dimension of this 'disrupted' reality, namely, that established agribusiness actors, including such 'heritage' companies as Bayer-Monsanto, John Deere, Dow-DuPont, and the Syngenta Group, are strengthening their hold over the agro-food system. As Bronson (2018: 11) observes, we need to be more aware of the 'productivist' value frameworks underlying PA technologies and "the ways in which these may be privileging already privileged actors in the food system". In other words, there are remarkably strong continuities between the corporate architecture of the 'old' industrial model and its extension in PA. These incumbent agribusiness conglomerates are clearly taking Coyle's (2018: 50) rhetorical question to heart: "Is it just a matter of time before platforms drive out incumbents with traditional business models, or before incumbents switch organisational model?"

Analysis and conclusion

The closer, targeted digital control over farm inputs is represented in some quarters as a new paradigm of 'sustainable intensification' that promises not only to raise productivity and farm profits but also to mitigate global climate change and help "feed the 9 billion" (Stanford Graduate School of Business, 2017; World Economic Forum, 2018). However, the 'sustainable' tag relates principally to variable rate applications of agro-chemicals, and input savings in some areas may well be offset by the need for higher applications elsewhere. More tellingly, as Gibson (2019: 159; my emphasis) cogently argues, variable input applications of plant protection chemicals and fertilizer are *discrete, reductionist interventions, "for which the market can supply remediation"*, fully in keeping with a market-based, business-as-usual scenario and continuation of the technology treadmill.

As such, these discrete, capital-intensive 'solutions' or disaggregated 'fixes' reproduce the dominant model of industrial agriculture and liberal market governance. In other words, PA technologies are not the forerunners of a coherently articulated, paradigmatic approach to sustainability transition (Gibson, 2019). For all the moral posturing on global warming and repetition of 'zombie statistics', like 'feeding the 9 billion', commercial gain is paramount. The sales pitch for Bayer-Monsanto's platform, Climate FieldView, makes this abundantly clear: the aim is "to optimise inputs to maximise yield and profitability on every acre" (cited by Rogers, 2018).

Critical social scientists regard the diffusion of PA technologies as opening up new circuits of value, extending corporate power, and accelerating technological competition (Carolan, 2017b), intensifying economic pressures on smaller farmers (IATP, 2020) and reinforcing land consolidation. Others see digitalisation as a vector of corporate control over farm work routines and production processes, eroding farmer autonomy, and adversely impacting rural labour markets and communities (Rose et al, 2018; Rotz et al, 2019a, 2019b). These and related issues of farmer identity, data ownership and privacy, knowledge politics, and ethics, are taken up in the next chapter.

3

Precision Agriculture: Adoption, 'Re-Scripting', Farmer Identity, Path Dependence, and 'Appropriationism 4.0'

In the previous chapter, we suggested that PA technologies and proprietary digital service platforms are intensifying the technology treadmill, reinforcing the dominant 'productivist' model and further extending corporate power over the agro-food system. This is convincing at a meta-level of analysis (see also Wolf and Buttel, 1996) but can give the false impression that technological innovation is an autonomous force in this socio-technical assemblage, uncontested by social actors, and fairly uniform in its consequences. A far more nuanced, less totalising view emerges when we examine the specificities of technology adoption and diffusion processes in US and EU agriculture and explore the contested issues these reveal.

Such specificities and contestations are better understood when seen against the backdrop of the long-term evolution of agrarian structures and their distinctive 'divides' and inequalities. At first sight, this may seem to be a lengthy, and somewhat unnecessary, detour. However, it highlights not only the socio-structural continuities of techno-scientific change over the past three decades but also the polarities it has produced and heightened. Attention to agrarian structures also provides a counterpoint to the micro-level analysis of innovation in PA that takes up most of this chapter.

Structural fault lines and change in US and EU agriculture: the last 30 years

Two well-turned phrases offer thumbnail synopses of the US agrarian 'condition' as depicted by metrics of concentration in farm size and farm incomes. The first of these phrases, revisiting its eponymous 1987 predecessor, is "Crisis by Design" (IATP, 2020). This report censures the last

30 years of farm policy *inter alia* for not alleviating the cost-price squeeze powering the 'technology treadmill' and farmer indebtedness, failing to articulate an effective response to climate change and extreme climate events, and neglecting the damaging effects of rising corporate concentration on farmers and rural communities.

The second phrase, 'Get big or get out', also figures prominently in the IATP report as a succinct summary of the structural inequalities exacerbated by the failings of US farm policy. This phrase, uttered in the 1970s by a former Secretary of Agriculture, Earl Butz, was loosely repeated in 2018 by Sonny Perdue, the Trump administration's Secretary of Agriculture. When questioned about the future of the dairy industry, he answered that "In America, the big get bigger and the small get out".[1,2] This statement, with its clear ideological resonance, is the best one-line summary of trends in farm size distribution and, by extension, farm population and rural social change since 1950 in North America and, for that matter, in the EU (see also van der Ploeg et al, 2015).

Or, that is, at least since Fred Buttel (1983: 92) referred to "the disappearing middle" of the "traditional independent, full time, medium-sized family farm", creating a dualistic or "bi-modal structure characterised by the increased dominance ... of extremely large farm units and by the increased prevalence of extremely small farms". Mid-sized cropland farms have continued to disappear in the US, consolidated mainly into larger units, so that the bi-modal size distribution has become even more pronounced in the intervening 30 years. The situation in the EU of 28 member states is complicated by the presence of nearly 4 million semi-subsistence farms in 2016 but the structure of commercial production is also marked by a bi-modal size distribution and consolidation. Small and medium-size farms, 68 per cent of all farms in the EU-28, produced 44 per cent of commercial output in 2016, while just 304,000 large farms accounted for the remaining 56 per cent (Eurostat, 2018).

Taking the last six census periods in the US, 1982 to 2012, the *average* size of cropland farms has changed relatively little but the *median* has grown steadily from 589 to 1,201 acres in the same period (MacDonald et al, 2018). "Consolidation in crop production is pronounced, nearly ubiquitous across commodities and states, and persistent over time" (MacDonald et al, 2018: 2). These authors also found that the share of cropland operated by large farms exceeding 2,000 acres rose from 15 per cent in 1987 to 36 per cent in 2012, with their numbers nearly doubling in that period, while those of mid-sized farms (between 100 and 999 acres) almost halved (27).

A further measure of farm size is annual gross cash farm income (GCFI), with small family farms defined as those earning under US$350,000 and large family farms as those whose GCFI exceeds US$1 million. In 2015, small family farms accounted for 90 per cent of all US farms but only 24 per cent of the value of production, whereas the same figures for large family farms were 2.9 per cent and 42 per cent, respectively (MacDonald and Hoppe, 2017).[3] The finding that in 2019, on average, these small family farms depended on off-farm sources for *over half of their total family income* underlines the polarity at the heart of rural economy and society in the US (Giri et al, 2021).

Continuing land consolidation and the pronounced bi-modal structure of farm size and income in US crop production over the past three decades emphasise the resource or wealth inequalities contextualising the diffusion of PA technologies. That is, only privileged members of the farming community have the financial leverage to afford the costly digitalised and automated equipment needed to generate the field data that delivers the economic benefits of variable rate input applications. Those small and mid-size conventional commodity producers who are on the 'wrong side' of this financial divide risk accumulating an unsustainable debt burden just to keep pace with the innovation treadmill, exacerbating the long-standing farm debt-income crisis in US agriculture and marking them out as eventual candidates for farm closure and land consolidation (Gloy and Widmar, 2014; Holtslander, 2015; IATP, 2020). In effect, with an R&D system dominated by corporate agribusiness, PA technologies are increasingly likely to exceed the financial reach of medium-size and small producers of conventional commodity field crops.

Livestock industries

The livestock sector offers a far more diverse picture of farm consolidation and "in one major sector – cattle-raising and its associated grazing land … [it] is yet to occur" (MacDonald et al, 2018: 2). Indeed, the share of US pasture and rangeland operated by farms exceeding 10,000 acres declined by 7 per cent in the years 1987–2012, while "twenty-eight million acres shifted to farms operating less than 500 acres" (MacDonald et al, 2018: 26). In other livestock sectors, with the exception of beef cow-calf operations, organisational changes, notably the rise of contract production and concentrated animal feeding operations (CAFOs) in the case of poultry, pigs, and cattle feedlots have brought pronounced increases in consolidation since the 1980s (Boyd, 2001; Watts, 2004; MacDonald et al, 2018; Howard, 2021).

In contrast to the relative stability of median herd size and farm numbers in beef cow-calf production during the past 30 years or so, the dairy sector has experienced dramatic and highly divisive structural change. In broad terms, the number of farms with milk cows fell from 200,000 in 1987 to 54,599 in 2017, although milk production is almost 50 per cent higher (MacDonald et al, 2020: iii).[4] These authors find that larger dairy farms have enjoyed marked long-term advantages over smaller operations, with lower production costs and higher net returns. A comparative study of dairy farms in the US and seven EU member countries also noted increased returns to scale and "an upward trend in farm net returns on assets with larger farm size" (Nehring et al, 2016: 228). Furthermore, milk prices in the US have become increasingly volatile since the early 2000s, notably in the crisis years of 2018–2019, when total costs exceeded gross returns and provoked a significant exodus from the industry, despite emergency federal government programmes (see also Mercier, 2019; Sharma, 2020). According to National Farmers Union data, "the average dairy farm has shown a positive net income only once in the last decade, in 2014" (Mercier, 2019).

As these figures suggest, this transformation has brought a significant surge in consolidation. Thus the median dairy herd size in the US rose steadily from 80 in 1987 to 1,300 cows by 2017, whereas the number of small, commercial family dairy farms with 10–199 cows declined dramatically from approximately 48,000 in 2007 to 30,000 in 2017 (MacDonald et al, 2020). Production has shifted decisively toward larger dairy farms, and those with over 1,000 cows now account for nearly half of the total US milking herd and more than 50 per cent of national output. In 2017, there were nearly 2,000 farms of this size, typically still family-owned, including so-called mega-dairies with over 25,000 cows, whereas, 25 years earlier, there were roughly only 500 such farms, which milked 10 per cent of all milk cows (MacDonald et al, 2020: iii).[5]

In the initial EU-10 member states, the pace of consolidation has been less pronounced, with more moderate changes and greater diversity in herd size, although mega-dairies operate in several countries. Nevertheless, and support under the EU's Common Agricultural Policy notwithstanding, the decline in the number of dairy farms has been remarkable: some *1.2 million farms, roughly four out of every five, disappeared between 1983 and 2013* (Augere-Granier, 2018). This striking decline occurred against a background of de-regulation in the EU dairy sector and, as in the US, rising concentration in downstream milk processing, manufacturing and retail outlets, depressing farm gate prices, often to levels below production costs, particularly over the last decade (Marsden et al, 2000, 2010; Howard, 2016, 2021; Sharma, 2020).

This brief discussion documents the inexorable rising tide of consolidation that has transformed agrarian structures in the US and the EU over the last 30 years. These growing inequalities are generating waves of farm closures as farmers are trapped by a cost-price squeeze and rising indebtedness. Indeed, the US has experienced rural de-population since 1920, when farm statistics were first collected, and this long-term rural exodus is broadly replicated after 1950 in Western Europe and Organisation for Economic Co-operation and Development (OECD) countries, particularly of secondary family members and non-family workers. Severe socio-economic consequences follow in the wake of these capitalist processes: declining rural communities with hollowed out social infrastructure and diminishing life chances (Gibson and Gray, 2019), and the toll of human lives. According to the Midwest Center of Investigative Reporting, 450 farmers committed suicide in nine Midwestern states between 2014 and 2018 (Hames, 2020).

Despite these harsh experiences, a common response is to look to innovation to raise operating net returns, exacerbating the deteriorating socio-economic situation even further. The brave new world of PA technologies now awaits, ranging from GPS-guided tractors and soil health monitors to robotic milking machines and the digitalisation of animal behaviour, offering a seductive though equivocal solution to these market pressures, as we see in the following sections.

Adoption and diffusion: a patchwork quilt

In the case of the US, the most thorough, wide-ranging study of adoption rates for PA technologies is based on the USDA's Agricultural Resource Management Survey of field crop production between 1996 and 2013 (Schimmelpfennig, 2016). Adoption rates for three technologies are reviewed: GPS-assisted mapping, including machine-mounted yield and soil monitors, GPS auto-steer systems, and VRT for input applications.

One of the central findings is what can be described as a *patchwork quilt* of adoption rates, with significant variations by farm size, type of technology, and crop. Measuring farm size by cropland area under cultivation, the highest adoption rates for all three technologies in 2010 were on farms exceeding 3,800 acres. The lowest rates were registered for VRT, the costliest of the three, on farms of all sizes. In general, non-adopters have much smaller farms than adopters – 480 acres smaller, on average (Schimmelpfennig, 2016: 17). "The largest corn farms, over 2,800 acres, have double the PA adoption rates of all farms: 70–80 per cent of large farms use mapping, about 80 per cent use guidance systems; and 30–40 per cent use VRT" (Schimmelpfennig, 2016: 1). These technologies were adopted on a higher proportion of the cultivated acreage devoted to corn and soybeans than for other field crops.

Schimmelpfennig (2016: 28) estimates that each of the three technologies has small positive effects on both net returns and operating profits: "the net returns of corn farms that use at least one of these technologies are 1.1 to 1.8 percentage points higher than for corn farms that do not use the technology ... mapping has lower capital requirements, on average, and the greatest impact on net returns (1.8 per cent)".[6] Although this study does not explore the socio-economic and political factors underlying the wide differences in adoption rates, these results suggest that the digital divide observed in PA is, at least in part, the expression of a financial and wealth divide. As Schimmelpfennig (2016: 13; my emphasis) comments, PA "equipment is highly specialised with limited resale potential and *is usually a sunk cost*", increasing the financial risks of adoption.

These findings resonate with those presented by Kernecker et al (2019) on the barriers to the adoption of smart farming technologies (SFT) as perceived by farmers and experts in seven European countries.[7] Both farmers and experts agreed that the high investment costs were the most significant barrier to adoption, reinforced by "a lack of clarity as to the added value that SFT would bring" (Kernecker et al, 2019: 10). These results replicate almost exactly those of an earlier, less extensive study of obstacles to SFT adoption in four Western European countries (Long et al, 2016). The study by Kernecker et al (2019: 11) also revealed a spatial dimension to the digital divide, noting that "Both farmer and expert groups from all sectors across Europe stated that poor broadband connectivity was an infrastructural barrier to SFT adoption". Similar infrastructural inequalities in connectivity also are a serious problem in rural North America (Janssen et al, 2017; Rotz et al, 2019a; Bloch, 2020).

As we have seen, long-term structural trends of land consolidation and specialised commodity production have persisted into the 21st century, but a more recent dimension is the "growing bifurcation" in the US between large-scale row crop producers and smaller farmers in organic agriculture and regional and local food systems (Dimitri and Hefland, 2020).[8] This bifurcation has coincided with public policy shifts in the management of market and production risks from agricultural price supports to crop insurance, *de facto* reliance on private technology development, and rising consumer awareness of food safety, health, and the environmental costs of intensive industrial agriculture (Dimitri and Hefland, 2020: 13). This marks a further 'divide' in the diffusion of PA technologies, effectively excluding smaller agro-ecological and bio-dynamic producers supplying regional and local food systems.

These technologies and smaller scale, low external-input production systems are not incompatible *per se* (Klerkx and Rose, 2020) but the current suite of PA innovations is closely tailored to conventional, large-scale, field crop systems, notably the Midwest corn/soya rotation, as Schimmelpfennig's (2016) findings clearly indicate. These are the 'agricultural futures' projected by the powerful corporations that dominate R&D in the industrial food system (see also Carbonell, 2016; Carolan, 2017a, 2017b, 2018; Bronson, 2018, 2019).

Although promising initiatives are emerging (Rotz et al, 2019a), the incorporation of 'alternative', diversified agro-ecological producers requires a radical sea-change in public policy to overcome the historical legacy of under-funding of organic agriculture (DeLonge et al, 2016), expand R&D in scale-appropriate digital technology (Winter et al, 2017), and increase public investment to extend digital infrastructure (Bronson and Knezevic, 2016).[9] Winter et al (2017) argue that, in principle, information and communications technologies can be used to promote self-organisation and farmer cooperation, including the development of platforms resembling Uber and Airbnb for the exchange and joint use of farm machinery. Again, however, such initiatives require supportive public policies to establish commercial and legal frameworks.

Re-scripting technologies and farmers: evidence from the UK

We have seen that the uneven, 'patchwork' distribution of PA technologies among US farmers revealed by Schimmelpfennig (2016) and the political economic factors behind differential adoption rates are broadly replicated in EU countries. However, some studies delve more deeply into the multi-faceted socio-economic factors underlying low or equivocal farmer engagement with these decision support tools (Wolfert, 2017; Klerkx et al, 2019). An insightful approach to the question of farmer ambivalence is taken by Rose et al (2018), who suggest that analyses so far have failed to ask *how* farmers use these technologies and to examine their impact on the *spatiality* of on-farm work and their *autonomy* in decision-making. This view is buttressed by evidence from questionnaire surveys and focus group discussions on the use of digital support tools by farmers and their technical advisors in three regions of England and Wales.[10]

To theorise the agency of farmers in using and shaping innovations, Rose et al (2018: 12) draw on Latour's (1992, 1993, 1994) concept of 'the script', which refers to the ways in which actions are mediated by artefacts, including technologies. In this light, "the use of digital support tools should be seen as a co-productive relationship between designers, knowledge brokers and end users; one in which tools are interpreted, resisted, and modified by users, whilst simultaneously re-scripting life on the farm." Combining this actor-network co-production framing with the 'social construction of technology' (Bijker, 1995) approach, Rose et al (2018: 13; my emphasis) suggest that "Users are ... able to transform technologies through resistance and negotiation, drawing on their own situated knowledge to *interpret technology in place. Farmers are rarely passive participants in farm innovation*".

These survey results illuminate factors behind the equivocal engagement of farmers with decision support tools, particularly the question of trust and their reliance on local knowledge networks and cultures. Thus, "The value placed on experience-based, situated knowledges was strong, which was the most significant reason why farmers (and advisors) did not use decision support tools or, if they did, why they failed to put complete trust in their recommendations" (Rose et al, 2018: 14). For many respondents, when compared to their long-standing advisors, "non-human decision support tools were considered to be placeless and thus to provide the view from nowhere ... far removed from the context of individual farms" (Rose et al,

2018: 15). In other words, "The perception that tools were not targeted towards individuals and instead were aimed at an imagined farmer, with a workflow and skills suited to their use, was one important reason behind the resistance, negotiation and ultimate re-scripting of tools by users" (Rose et al, 2018: 15).

Rose et al (2018) also explore the ways in which decision support tools have, in turn, re-scripted end users themselves by requiring new spatial patterns of work, notably office-based decision-making, which conflict with their habitual outdoor styles of management. Significantly, respondents emphasised that these work patterns weakened their relation to the land, the bedrock of their identity as a farmer. For some farmers, changes in their material relationship to the farmed landscape would "negatively impact upon their quality of life on the farm" (Rose et al, 2018: 16).

This prompts Rose et al (2018: 16) to argue that the re-scripting of spatial workflows encroaches on the imagined space of farming held by individual farmers in ways that may "lead to the re-scripting of agricultural society (or at least its imaginaries) at wider scales". In common with other studies of the relation between digital innovations and farmer subjectivities, Rose et al (2018) identify:

> a mismatch between what farmers imagine farming to be, and the imagined space of farm decision-making as conceived by developers and proponents of technology. In other testimonies, there was a clear tension between some farmers' understanding of their enterprise, and the perceived direction of travel of the industry as imagined by these farmers. (Rose et al, 2018: 17)

In the study of perceptions of SFT in seven European countries discussed earlier (Kernecker et al, 2019), experts also noted the tension created by the digitalisation of farm work processes and the changing profile of farmers. They are convinced that "Farmers will function more as managers and supervisors of machinery instead of actually working in the fields. The next generation of farmers are expected to somehow naturally adopt SFT as they are "sons of the internet" (Kernecker et al, 2019: 9).

Farmer identity, autonomy, and control

This discussion reveals the complex dynamics of corporate power and farmer subjectivities that permeate digitally mediated changes in the material nature and spatiality of farm workflows, and the shifting balance between tacit and digital knowledge, and how these affect perceptions of the 'good farmer'. In this section, we focus on how digitalisation is reconfiguring farmer identity, autonomy, and control, as revealed by hotly contested issues of data ownership, privacy, and knowledge production.

In some ways, following Rose et al (2018), the question of farmer identity and autonomy is relatively straightforward, insofar as reliance on digital decision tools inevitably and inexorably changes how the farmer understands *the materiality and spatiality of the farm*. Without these tools, this understanding is largely a matter of personal observation and experience. Contrast this process of knowledge production with the farmer's role on a highly automated, large-scale arable farm, where "mobile digital technologies and analytical software packages have transformed the tacit and embodied knowledge of the farmer (their 'feeling' for [the] land, one might say) into quantified automated procedures, using digital data that is captured largely autonomously and processed algorithmically to give actionable spatial knowledge" (Dodge, 2019: 39).

Admittedly, this statement describes the high-tech end of the automated production spectrum since, in practice, as we know, PA technologies are adopted piecemeal and often 're-scripted' (Schimmelpfennig, 2016; Rose et al, 2018). Nevertheless, it encapsulates the concern that farmers' professional knowledge and skills are being superseded insofar as "Code has traduced farmers into screen-workers" (Dodge, 2019: 41). Of course, as the experts interviewed by Kernecker et al (2019) recognised, this could be a generational problem: the "sons of the internet" will take code and screen time in their stride and forge new farmer identities steeped in digital production practices and so redefine the notion of the 'good farmer'.

This glib comment has intuitive appeal but obviously it elides the subjective costs and 'collateral damage' that this transition imposes on the present generation of farmers. Jane Gibson (2019), for example, argues that "Now, not just machines, but new kinds of experts stand between farmer and land" as "natural systems are rendered as binary code". This shift devalues hands-on, experiential knowledge of the land and "involves *a restructuring and relocation of power* … in favour of data captured, distributed and analysed by sophisticated software" (Gibson, 2019: 25; my emphasis).

Such digital intermediation raises the issue of "whether large-scale industrial farmers, whose decisions may be influenced or even made by IT specialists, will be able to hold onto any long-run, nonmarket concerns for social and environmental health" (Gibson, 2019: 27). The risk that stewardship of the land and other valued attributes of 'good' farming practice are in danger of being undermined is clearly articulated by one of Michael Carolan's (2017b: 11) respondents:

> Yields are great but I worry about how technologies like this distract from those other things we're growing – biodiversity, trust, strong communities. If we all start evaluating each other based on what we're

hauling to the elevator every fall, that's not the culture that attracted me to farming (Paul, farmer).

The "restructuring and relocation of power" accompanying digitalisation goes to the heart of farmer identity and autonomy issues. In the absence of robust legal and regulatory frameworks, farmers adopting PA technologies effectively surrender control over farm data to their corporate technology providers (Wiseman et al, 2019). These companies "have introduced lengthy and complex software licence agreements that govern the way that farmers' data will be collected, managed and shared with their smart farming technology providers" (Wiseman et al, 2019: 2). The loss of control over the way farm data are managed and the lack of trust that technology providers will respect farmers' privacy, not surprisingly, generate 'mixed feelings' about the adoption of these technologies (Wiseman et al, 2019: 10).

At the same time, analysing the results of a detailed survey of 1,000 Australian farmers across 17 sectors, Wiseman et al (2019: 2) find that farmers frequently are unaware "of the scope and extent of the terms of the software licences embedded into farming equipment (e.g. the sensors, robotics, drones, tractors, and the agricultural machinery)". This relocation of power is nakedly exposed by the fact that "the mere act of turning on their machinery or downloading the technology means that they have agreed to a broad range of terms that regulate who can use and access the data generated on their farm" (Wiseman et al, 2019: 2). Consent to manage and use on-farm data thus is ceded to competing corporate technology providers, each with a proprietary smart farming platform ring-fenced by intellectual property rights that lock "farmers into a specific brand and operating system" (Rotz et al, 2019a: 211). In effect, farm data is accumulated in corporate silos, adding a further dimension to vertical integration and accentuating agribusiness concentration.

The 'right to repair' controversy is frequently taken as the consummate expression of the loss of property rights and disempowerment of farmers by PA technologies (see also Plume, 2014a; Sykuta, 2016). In the words of Kyle Wiens (2015), founder of iFixit, an online DIY community,

> there's not much you can do with modern ag equipment. When it breaks down or needs maintenance, farmers are dependent on dealers or technicians – a hard pill to swallow for farmers who've been maintaining their own equipment since the plow ... the problem is that farmers essentially are driving around a giant black box ... [and] only the manufacturers have the keys to those boxes ... John Deere, in particular, has been incredibly effective in limiting access to its diagnostic software ... The dealer-repair game is just too lucrative to cede control back to farmers.[11] (Wiens, 2015)

This dispute is taken by Carolan (2018) to demonstrate the ambiguity of the concept of 'access' outside its specific political economic context. "To put it plainly", the terms of the 1998 (US) Digital Millennium Copyright Act (DMCA), "made it illegal for a tractor's owner to access the 'brains' of their smart equipment – that is, the engine control unit" (Carolan, 2018: 749). In 2015, an exemption to the DMCA was granted to allow individual owners, and these alone, to 'tinker' with the code. This was a Pyrrhic victory, however, since "as most farmers lack specialised training in code and software, this narrowly granted exemption does nothing in practice to reduce their dependency on agro-food firms and 'approved' technicians. The exemption is an example of how … 'free access is not necessarily fair access'" (Carolan, 2018: 755).

Lacking an effective regulatory architecture, a variety of farmer cooperation and self-organisation initiatives have emerged to provide alternatives to the insecurity of data ownership and address privacy concerns associated with proprietary decision support platforms and resist the dependence and disempowerment they engender. That is, to reconfigure how digitally mediated knowledge is produced, by whom, and in whose interests. These include Farm Hack, an online community of farmers formed in 2011 to "build and modify our own tools", supported by online documentation on how to repair analogue equipment (Carolan, 2017a; Cosgrove, 2017; Farm Hack, 2018).[12] More recently, Farm Hack has developed farmOS, an open source platform with custom modules that can be used for a range of different production systems.[13] In contrast to the lack of interoperability between rival corporate platforms, Don Cox, a co-founder of Farm Hack, notes that "in building a common system, we're able to make some of the decision tools that come out of academia or other agricultural organisations … compatible with one another". With this system, "The fundamental things are that farmers can enter their data once and they have control over that data in a secure server. It's not in the cloud somewhere" (Cox, cited by Cosgrove, 2017).

Farm Hack certainly is not alone in contesting corporate domination of the politics of knowledge production, which places farmers on the weaker side of the 'digital divide'. Other exemplars are the Open Ag Data Alliance (OADA), in association with Purdue University, and the Global Open Data for Agriculture and Nutrition initiative (GODAN). Farmer data rights also are championed by some farm management start-ups, such as Farmobile, that wants to persuade farmers to treat their data "as another crop, which they can sell or licence to provide a permanent revenue stream" (Poppe, 2016; see also Bloch, 2019). While these efforts point to alternative paths of public policy and governance, Carolan's (2018) point that access should not be confused with capability, autonomy and fairness is still supremely relevant. As put by Rotz et al (2019a: 219–20), farmers "are not *necessarily* any less dependent on the physical and technological products and services that are embedded in corporate assemblages ... and being developed by an increasingly consolidated and corporately controlled agricultural industry" (original emphasis).

SFT and path dependence

As we have noted previously, *enclosure* is a key competitive strategy of 'platform 'capitalism', funnelling customers and their data into 'silos' or closed proprietary digital 'ecosystems', leading to user dependence and lock-in (Srnicek, 2017). The International Panel of Experts on Sustainable Food Systems (IPES-Food, 2017) shares this view, identifying Big Data technologies as a major new driver of consolidation, and arguing that these innovations are reinforcing rather than disrupting the incumbent regime of farming and food production. To support this conclusion, IPES-Food (2017: 56) drew attention to the parallel concentration of R&D expenditures, which is creating "path dependencies by focusing innovation on a narrow range of crops, technologies and approaches", to the detriment of genetic innovation in staple food crops produced in the Global South. Indeed, according to Mooney (2015: 119–20) seed companies concentrate "almost all (their) investment ... in no more than a dozen crops", and "45% of global private investment is on one crop – maize." In response, IPES-Food (2017: 86) advocate a "change of paradigm" from the "current high-tech approach that governs knowledge and innovation" to a "wider-tech paradigm that would shift the focus to diversified and decentralised innovation, locally applicable knowledge, and open access".

In the lexicon of evolutionary economics, path dependencies, as in the trajectory of digital agricultural innovation, are the result of processes of technological, cognitive, epistemic, and institutional lock-in that entrench the paradigm of productivist, industrial agriculture (Arthur, 1989; Vanloqueren and Baret, 2009; Bruce and Spinardi, 2018). In other words, PA technologies are embedded in particular forms of techno-scientific knowledge and design practices, which privilege certain styles of farm management and modes of farm livelihood. For example, reviewing the digital marketing campaigns of John Deere and Monsanto, Bronson and Knezevic (2016: 3) ask "in what way do the images circulating in the promotion of Big Data tools normalise hegemonic farming systems?"

Just as we speak of technological lock-in and path dependence, so particular social relations, forms of productive organisation, and power structures are reproduced by these innovations. This position recalls neo-Marxist formulations of the 1970s and 1980s, now rarely recognised, that capitalist social relations are imbricated in the science and design of technology, which reproduces established patterns of power, class exploitation and value capture (see also Young, 1979; Levidow and Young, 1985).[14]

This theme has been re-visited by Kelly Bronson (2018, 2019) in interviews with scientists, engineers, and policymakers involved in agricultural innovation in North America. She concludes that they hold a narrow set of values and 'productivist' imaginaries of a 'good farmer' and 'good technology', with a focus on maximising yields, that bias technology design towards large-scale, mono-crop production of staple commodities, often destined for global markets. As a section heading has it, "When old is new: smart agriculture's value frameworks" (Bronson, 2018: 9). This is also illustrated by:

> a corporate video – *Farm Forward* – that is meant to project John Deere's vision ... for its PA equipment ... The American farmer of the future – like 'Terry' in the campaign video – is able to farm from the comfort of his living room, where these tools give him a god's eye view of his fields [translated into data points]. (Bronson, 2018: 9)

As she observes, farm labourers displaced by automation are conspicuous by their absence from this picture.

This unequal political economy is stimulating a diverse body of literature in critical agro-food studies that is engaging with the agricultural knowledge and innovation system (AKIS) and issues of lock-in and dependence. One strand, briefly mentioned earlier, addresses the neglect of bottom-up, farmer-based, scale-appropriate digital technologies (Winter et al, 2017; Rotz et al, 2019a). Related patterns of inclusion/exclusion have prompted a recent 'turn' to the notion of 'responsible innovation' to give voice to actors typically excluded from technology decision-making and to address ethical concerns (Bronson, 2018; Rose and Chivers, 2018; Klerkx and Rose, 2020). Taking another tack, a series of papers by Michael Carolan (2018, 2019, 2020, 2022) articulates a performative approach that focuses on the effects technology engenders and the embedded political ontology of market-led individualism and neoliberal governance propagated by top-down corporate PA. In a nutshell, this political ontology creates the conditions for its realisation. Frequently mentioned but under-researched themes are the dis-equalising effects of these technologies on rural labour markets[15] (Rotz et al, 2019b) and the socio-economic viability of rural communities (Gibson and Gray, 2019).

'Appropriationism 4.0': analysis and conclusion

Over 30 years ago, my colleagues and I conceptualised the industrial transformation of agriculture in the late 19th and 20th centuries in terms of the capitalist appropriation of on-farm activities, their commodification as industrial inputs, and re-integration in the farm production process

(Goodman et al, 1987). We analysed this transformation on two interrelated fronts: first, mechanisation and the transition in the energy economy of agriculture from biomass and animal power to fossil fuels, exemplified by the tractor, combine harvester, and ever-greater reliance on synthetic agro-chemicals. Second, the corporate appropriation of plant biology, marked by the passage from farmer seed saving and exchange networks, hand pollination, and varietal development by public agricultural research stations to the privatisation of the seed, initially via the 'biological lock' of hybrid vigour, the consolidation of the seed industry and the introduction of proprietary biotechnologies in the 1990s.

Then as now, these socio-technical transitions reconfigured knowledge production: the situated knowledge of managing draught animals and seed saving was marginalised – "a collective forgetting" in Carolan's (2017a) evocative phrase – and farmers were compelled to acquire new skills. The current diffusion of digital technologies is re-scripting daily lives, work and occupational space-time patterns across farm landscapes as tacit knowledge based on qualitative assessment, insights, and observation is displaced and translated into code (Rose et al, 2018; Dodge, 2019; Gibson, 2019). Local, *lived* geographies give way to the abstract assemblages of algorithmic rationality. More generally, Mau (2019: 13) refers to quantification as "a process of 'dis-embedding', which deliberately strips away local knowledge and the context of social practices in order to obtain more abstract information that can be combined ... with information from other sources."[16]

This transition from field-scale to sub-field scale management involves not just the 'dis-embedding' of farmer knowledge and authority, but their transformation from a use-value into a monetised resource. In a classic Appropriationist dynamic, experiential knowledge is re-integrated into the production process in the newly commodified form of corporate decision support services and input 'prescriptions'. As Miles (2019: 8) notes, "Digitalisation provides an opportunity to capture, access, manage and exchange what was inaccessible before".

As in earlier periods, the conquest and 'colonisation' of landed space again is central to Appropriationism 4.0. However, the socio-technical assemblage of PA is now extending the commodity form into new spheres of the production process by *miniaturising* the farm landscape.[17] In the 20th century, the appropriation of agricultural space occurred by extension, with mechanisation displacing farm labour, accelerating rural de-population and land consolidation: "It was the mobile machine, the tractor, which came to symbolise labour appropriation" (Goodman et al, 1987: 120). Today, industrial appropriation is enacted by *reducing* extensive space to digitally demarcated, high-resolution, sub-field management zones – a 'digital land grab' (Mau, 2019: 21). In the words of the Bayer-Monsanto Climate FieldView platform, it will treat "every acre with variable rate seeding prescription tools, nitrogen management tools, and fertility scripting tools" (cited by Rogers, 2018).

This is a new expression of power that represents agricultural space "in ways that *privilege abstraction and calculability*" (Ash et al, 2016: 3; my emphasis), normalising and deepening its commodification by corporate capital. The tractor and combine harvester are again restructuring farm production only now, armed with sensors, spatial imaging devices and integrated software systems, they are harvesting pixels to serve a new algorithmic logic of capital accumulation.

As Dodge (2019: 41–2) remarks, "The code underpinning PA and the algorithms in expensive machinery like combine harvesters have developed to a point where there are attempts at a fully automated arable production system using smart technologies, Big Data and machine-learning algorithms." In this pre-figurative, 'Appropriationist' vision, robotic, factory-like production will accelerate the loss of farm families and rural jobs, condemning rural communities to a moribund future.

This theme of miniaturisation and the extension of the commodity form is carried forward to the next chapter on the development of alternative plant-based and lab-engineered proteins. This rapidly expanding assemblage exemplifies the convergence of ICT, biotechnology, and synthetic biology in the drive to identify and combine nutritious plant compounds to replace farmed animal protein, and represents the contemporary expression of Substitutionism.

4

Alternative Proteins: Bio-Mimicry, Structuring the New Protein Industry, 'Promissory Narratives', and 'Substitutionism 4.0'

The alternative proteins (AP) industry offers further evidence of the fertile convergence between software-enabled ICTs and DNA-enabled biotechnologies that underlies structural change in modern agro-food systems. The myriad inter-related disciplines engendered by this convergence, such as genome sequencing, bioinformatics, synthetic biology, and cellular engineering, are at the heart of this nascent industry, aka 'molecular farming'. Advocates claim that this emerging sector will contribute significantly to environmental and human health and animal welfare by substituting alternative proteins for conventional farmed sources, structurally undermining the livestock-feed grains complex, the fulcrum of today's industrial food system, and a major source of GHG emissions.

Technological miniaturisation and reductionism again are central mechanisms underpinning these transformative claims since the aim is to replicate meat, fish, milk, and eggs by identifying their constituent properties at the molecular and cellular level. These mechanisms are pinpointed in the next section, which reviews the innovative technologies developed by the pioneering start-up firms in the two main branches of the AP industry: plant-based meat analogues and cellular agriculture, comprising tissue culture engineering and fermentation-based protein production systems.[1] We then focus on the emerging industrial structure of the AP space as Big Food actors are drawn to this rapidly growing sector and launch their self-styled future as 'protein corporations'. The penultimate section explores the contested ontological politics of the discursive strategies and 'promissory narratives' deployed by AP start-ups and Silicon Valley venture capitalists, and particularly the claim that the new industry is the forerunner of the paradigm shift needed to meet the multiple challenges of global climate change. A final section reflects briefly on the 'substitutionist turn' or Substitutionism 4.0 as the protein economy diversifies away from its traditional roots.

Bio-mimicry and its technologies
Plant-based protein

The use of microbial fermentation to produce non-animal proteins, particularly with yeasts, dates back to the ancient practices of the baking, brewing, wine-making, and food industries. Jumping forward to the post-Second World War years, advances in chemical and biochemical engineering raised the efficiency of whole cell and enzyme fermentation processes and extended the range of feedstocks that could be used. However, the real breakthrough came with the application of recombinant-DNA methods in industrial microbiology in the later 1970s and 1980s (CTA, 1981, 1984). These powerful new tools, reinforced by protein engineering[2] and later rounds of biotechnological innovation, paved the way for the eventual rise of Beyond Meat, Impossible Foods, EAT JUST and other start-ups of this ambitious AP cohort.[3]

Among their precursors, single cell (microbial) protein (SCP) briefly raised the prospect in the 1960s of producing alternative proteins commercially from both renewable and hydrocarbon feedstocks (Litchfield, 1983). However, for a variety of reasons, including the Organization of the Petroleum Exporting Countries (OPEC) oil price shocks in the 1970s, these prospects faded, and hopes for SCP were then vested in mycoprotein, a micro-fungus derived from the soil mould *Fusarium venenatum* and produced by fermentation on starch feedstocks (Bud, 1994; Sexton, 2014). Its most successful exemplar, known by its brand name, Quorn, was launched in 1985 by Sainsbury's as the main ingredient of savoury pies but failed to break out from its market niche into the mainstream until the current upsurge in demand for alternative protein foodstuffs.[4,5]

With this brief historical background, we turn to contemporary plant-based meat substitutes which, by achieving bio-mimicry or 'visceral equivalence' with conventional farmed meat, are successfully entering the mainstream (Stephens et al, 2018; Santo et al, 2020). Replication of the molecular properties of farmed meat involves the use of robotics and machine-learning to screen vast proprietary plant databanks to select promising proteins and compounds for predictive modelling before scaling up for fermentation in bioreactors (Poinski, 2019).[6] As Bradshaw (2014) puts it, "Fed with enough data, the system can … identify relationships between molecular properties, such as weight and a particular functionality, that make food taste just right", adding that "A proprietary algorithm is the magic ingredient of many a Silicon Valley start-up rather than one related to food!". The nexus between the information and molecular sciences that underpins plant-based protein technologies is captured by a venture capitalist's description of Hampton

Creek Foods, a company now known as EAT JUST, maker of plant-based egg substitutes, Just Mayo, and liquid egg, Just Egg: it is "that rare thing, a company at the intersection of machine learning and plant biological properties, and branding" (cited by Bradshaw, 2014).

The proteins in plant-based meats are obtained primarily from legumes and cereals in whole or fractionated form as flour, concentrates, isolates, and hydrolysed isolates (GFI, 2021)[7] These products can involve one or a combination of ingredients obtained from plants, fungi, and microbial cell culture, including proteins produced by recombinant DNA engineering methods. The textured vegetable proteins are processed mainly by extrusion techniques, but recent innovations include sheer-cell technology and 3D printing methods. Several plant-based start-ups, such as the Spanish firm, Novameat, are developing 3D or 'bio-printing' and the fine micro-extrusion methods needed to produce more highly textured meat cuts, like steak. This company hopes eventually to license its patented technology to established food manufacturers (Carrington, 2020).

The Good Food Institute (GFI, 2021) also reports that some start-ups are eschewing mechanical methods altogether in favour of biological techniques, such as microbial fermentation. It suggests that Perfect Day's ice cream made with recombinant casein proteins gives a "preview of how fermentation-derived ingredients can provide plant-based meat, eggs and dairy with the organoleptic properties of their animal-based counterparts" (GFI, 2021: 22).

Cellular agriculture: tissue engineering systems and fermentation-based meat substitutes

Cultured meat, aka 'clean' meat and cell-based meat, is produced by tissue culture engineering using stem cells taken by biopsy procedures from adult live 'donor' animals, which are then grown *in vitro* and differentiated into muscle cells before these are incubated "in tank-like bioreactors in a soup of proteins, sugars, and vitamins" (Kazan, 2019).[8]

Fassler (2018) recounts a visit to EAT JUST in San Francisco and being shown a video of 'Ian', a chicken still living in the nearby Santa Cruz Mountains, whose stem cells were being used to produce tissue-cultured chicken nuggets. For this observer, 'clean meat' is "A near-future brave new food, more dependent on AI than the physical needs and limitations of animals." Rather than depend on a constant supply of biopsy-derived stem cells, a genetically modified cell line can be produced, which only involves animals initially in order "to source the original cells" (Stephens et al, 2018: 157). However, the technology of this second pathway is new and raises serious safety issues and other regulatory concerns (Santo et al, 2020).

'Bio-printing' or 3D techniques are also used to produce tissue engineering-based meats, and an Israeli company, Aleph Farms, in conjunction with biomedical engineering researchers at the Technion-Israel Institute of Technology, recently "announced that they have created the world's first slaughter-free rib eye steak" and hope to begin marketing its products in 2022 (Poinski, 2021a). A second Israeli cultured meat firm, Meat-Tech 3D, also uses bio-printing methods and reportedly is preparing for an Initial Public Offering (IPO) in the US (Poinski, 2021b).

In contrast to tissue culture-engineering systems, fermentation-based alternative proteins do not require stem cells from living animals and are produced in bioreactors using cells from "bacteria, algae or yeast that typically have been genetically modified by adding recombinant DNA, so they produce organic molecules" (Stephens et al, 2018: 157). Drawing on such "commonplace industrial biotechnologies", "These molecules can be used to bio-fabricate familiar animal products", such as gelatine, casein, and collagen (Stephens et al, 2018: 157). Start-ups in the precision fermentation branch of cellular agriculture include the EVERY company (formerly Clara Foods), which recently introduced lab-cultured egg white, Modern Meadow, using genetically engineered cells to produce collagen for leather products, and Perfect Day, which produces whey protein for its animal-free product line of ice cream, feta, cream cheese, and spreads. In June 2022, in partnership with Mars, this company launched a dairy-free milk chocolate bar, CO2COA, based on its fermented GE dairy proteins (Poinski, 2022d).

Corporate routes to mainstream markets

We have already emphasised that each AP firm aims to mimic perfectly the sensory properties – taste, texture, mouth feel – of the animal protein product – meat, dairy, egg white – that it is seeking to displace. This *visceral mimicry* is vital to their commercial ambitions: the plant-based substitutes now widening their markets must be indistinguishable from the 'real thing' if they are to compete equally for space on supermarket shelves, fast food counters, and as ingredients in food manufacturing.

This almost obsessive concern for an identical match emerges clearly in a statement by Pat Brown, CEO of Impossible Foods:

> We describe our flagship product, the Impossible Burger, as 'plant-based meat' because the Burger's key ingredient [the soy protein, 'heme'] is the exact same ingredient found in animal-derived meat. In similar concentrations, with a similar nutritional profile, resulting in a similar experience for people who eat it – but our product contains no animals whatsoever.
>
> [...]

> The key ingredient is soy legehemoglobin … a protein that carries heme, an iron-containing molecule that occurs naturally in every animal and plant. Heme is an essential molecular building block of life, one of nature's most ubiquitous molecules. … [and] is super-abundant in animal muscle. It is the abundance of heme that makes meat uniquely delicious. (Cited in FAIRR, 2019: 15)

While the Impossible Burger and Beyond Meat's burger both rely on such plant-based ingredients as wheat protein, potato protein, and coconut fat, in other ways, Impossible Foods is an outlier. That is, insofar as it uses genetic engineering to produce the protein, soy leghemoglobin or 'heme', which, according to the company website, "is the molecule that makes meat taste like meat" and makes its eponymous burger 'bleed' like ground beef. The company scanned and analysed massive numbers of plant proteins and compounds before discovering a molecular substitute for animal heme in the root nodules of the soya plant. However, for production purposes, heme is obtained by inserting the gene for leghemoglobin into a genetically modified yeast and produced through cell culture and fermentation.[9] The animal-free milk company, Perfect Day, has broadly taken the same path, engineering the genes for the principal milk proteins, casein and whey, into the cells of a natural microflora, *Trichoderma*, which are then grown in fermentation tanks and harvested for transformation into finished products.

Cellular meat products are at a much earlier stage of development and face considerable scepticism about the timespan required to move from niche to commercial scale and price parity with farmed animal meat. Nevertheless, outgrowing its academic origins in biomedical science, tissue-culture engineering systems are now an active sector in terms of start-ups, venture capital funding, M&As, and investment by major agribusiness and food corporations seeking to build their 'protein portfolios'. This interest is a vote of confidence by investors that the sector can overcome its dependence on inputs that are currently derived from farmed animals: collagen-based scaffolds on which muscle tissue is grown (Stephens et al, 2018) and, most significantly, cell-growth serum, which is a major production cost with a volatile supply chain.

Although several firms have claimed success recently, as we will note, commercial expansion has been particularly inhibited by the lack of an animal-free growth medium. At present, this vital and very costly ingredient is "a blend of growth-inducing proteins usually made from animal blood. The most common is foetal bovine serum (FBS), a mixture harvested from the blood of foetuses excised from pregnant cows slaughtered in the dairy and meat industries" (Reynolds, 2018). In addition to supply chain problems, this continued reliance on farmed livestock raises animal welfare issues and badly tarnishes its commercial prospects. Santo et al (2020: 13) cite an estimate by Jochems et al (2002) "that 800,000 litres of FBS were produced annually worldwide for use in culture media, including for pharmaceuticals, corresponding to the foetuses of about two million cows".

Not surprisingly, these cost and marketing concerns have prompted sustained research efforts to develop a plant-based substitute for FBS. According to press reports on the regulatory approval of EAT JUST's tissue cultured "chicken bites" for sale in Singapore, this company announced that a plant-based serum will be used in the future. At present, this new product is far from being cost-competitive, will be sold in only one restaurant initially, and faces a daunting challenge to scale up production – "we need to move to 10,000 litres or 50,000 litre-plus bioreactors" (Josh Terrick, CEO, EAT JUST).

After years of R&D investment, Poinski (2021d) reports that a number of companies, including Mosa Meat, Future Meat Technologies, and Upside Foods (formerly Memphis Meats), are now poised to introduce plant-based alternatives to FBS. Although this breakthrough reportedly will reduce production costs massively, scaling problems remain, and cell-cultured meat products have yet to gain regulatory approval for sale in the US and other major markets. In these circumstances, it is hard to disagree that tissue engineering-based meats "remain for the most part at the prototype stage of development" (Santo et al, 2020) and are very much the junior partner in the 'alt-protein' industry at this point.

The future development of tissue engineered meat will depend on its success in making the transition from processed minced meat products – burgers, meatballs, and sausages – to whole cuts of meat, such as chicken breasts and beef steaks (Froggatt and Wellesley, 2019).

> The technical process involved in producing a steak *in vitro*, for example, requires culturing a more complex tissue, including multiple cell types, and considerable progress is needed to achieve a steak or whole-cut of meat that achieves the colour, flavour, and nutritional profile of meat harvested from an animal – and to do so in a manner that is economically viable is even more challenging, and therefore significantly further from the market. (Froggatt and Wellesley, 2019: 13)

The cultured or 'clean' meat industry has made great strides in the past decade, exemplified by Mosa Meat, Aleph Farms and Meat-Tech 3D, but the scale of this further transition tempers hopes in its potential to mitigate the multiple, over-lapping crises of global warming, as we discuss more fully later on.

Structuring the alternative protein industry

The origins of the AP industry are to be found in Silicon Valley venture capital funds and high-tech start-up firms deploying Big Data analytics, recombinant DNA techniques, and synthetic biology to produce animal-free proteins. In this emergent stage, online 'foodtech' sources report extensively on new start-ups, funding rounds and their changing cast of investors.[10] We begin, however, by examining the business strategies of the more established start-ups, including Beyond Meat, whose IPO on the New York Stock Exchange in May, 2019 was heavily over-subscribed, reaching a record-breaking market capitalisation of US$3.9 billion.

The overriding aim of these strategies is to reach mainstream scale, establish brand recognition, and acquire market share as quickly as possible by expanding their own production capacity, and/or partnering with larger food companies to achieve the requisite size. This scaling-up phase has become increasingly challenging as Big Food corporations are now launching their own products in the alternative protein market. The race to gain a firm foothold in mainstream retail outlets before these mega-corporations become dominant players is exemplified by Beyond Meat and Impossible Foods, each of which had to overcome initial supply difficulties.[11] The former has reached sales agreements with Whole Foods, Target, and other retailers, and Carl's Jr. fast food outlets, as well as Sysco, the leading food service distributor in the US. Likewise, Impossible Foods, once it had cleared the regulatory hurdles raised by its use of genetically modified material in the production of 'heme', is selling its burgers in US grocery stores nationwide, as well as in fast food chains, such as White Castle and Burger King, while some Starbucks cafes are featuring the Impossible Whopper and its sausages on the breakfast menu.

Beyond Meat consolidated its position impressively in 2021, announcing two strategic global distribution agreements. First, with McDonald's to supply the patty in the McPlant, a new plant-based burger being trialled in several international markets, and to create other plant-based menu offerings. A second global agreement is with Yum Brands to "develop a range of exclusive plant-based protein menu items for (its) KFC, Pizza Hut and Taco Bell chains" (Ellis, 2021a). In the words of the executive director of the Good Food Institute, a leading booster of alternative proteins, "With more restaurants and revenue than any other food chains on the planet, McDonald's and Yum Brands will bring plant-based meat onto the mainstream menus of millions of people. When these restaurant chains move, the entire food industry takes notice."

EAT JUST has taken a different course and is licencing companies in the US, the EU, and Asia to produce and distribute its plant-based egg products, while holding technical patents on its production processes. In this way,

EAT JUST builds on the production, distribution, and marketing expertise of its partners to reach commercial scale. Perfect Day has followed a variant of this path by partnering with Archer-Daniel-Midland (ADM), a leading global food and feed processor, to utilise its fermentation capacity to produce animal-free whey protein as an ingredient for a variety of food products. Perfect Day expects its whey protein to match the cost of conventionally sourced protein in a few years' time. ADM further extended its interests in supplying alternative proteins to food and feed companies by reaching an agreement in November, 2020 with the French firm, InnovaFeed, to build a plant in Illinois to produce insect protein for animal feed and other uses.

These steps by ADM and its partners indicate that the AP sector is acquiring a more mature industrial profile with the emergence of an infrastructure of intermediate suppliers of protein inputs for firms downstream. By deploying synthetic biology techniques of molecular and cell engineering and microbial fermentation,[12] firms such as Gingko BioWorks and its spinout, Motif FoodWorks, Nutriati, Amyris, Benson Hill, and Myco Tech, are designing and manufacturing customised microbial inputs for a range of industries, including flavours, cosmetics, health, food, and nutrition. The mission of Motif FoodWorks, for example, is to supply bio-engineered ingredients to alternative protein food companies to improve and diversify their product lines. A further sign of growing maturity is the formation of plant-based food associations in the US and the EU to lobby on regulatory questions and other governance issues.

Enter the 'protein conglomerates'

In response to this growth and re-structuring of protein markets, Big Food companies have been active in the seed funding rounds of start-up firms, in reaching partnership agreements, and making outright acquisitions in both the plant-based and cellular agriculture sectors. The power of M&A as an entry strategy and mechanism of concentration was vividly demonstrated once again in 2017 with Danone's US$12.5 billion takeover of WhiteWave Foods, producer of the leading ranges of plant-based dairy substitutes, *Alpro* and *Silk*, formerly owned by the Dean Foods dairy conglomerate. Danone has since accelerated its diversification, taking over a succession of companies producing a variety of plant-based foods, including creamers, cheese, and probiotics. As the case of Danone illustrates, this traditional strategy holds the key to the rapid response by the major animal protein corporations to the rising consumption of alternative protein products.

However, if we were to relate the many actors and the different forms of involvement between start-ups and established firms, not to mention the activities of 'accelerator' funds, 'incubator' programmes and the venture capital arms of large corporations,[13] we would be overwhelmed by detail. For example, in 2019 alone, global investment in agri-foodtech start-ups reached US$19.8 billion, encompassing 1,858 deals and 2,344 individual investors (Martyn-Hemphill, 2020). In addition to the constant threat of M&As and other entry strategies, the start-up firms now reaching commercial scale arguably are just as vulnerable, if not more so, to *direct* sales competition from Big Food firms with their large production and R&D facilities, supported by well-established supply chains, distribution, and marketing systems and, in most cases, global reach.

Since Beyond Meat's IPO in May, 2019, a roll call of Big Food and Big Meat corporations, including Tyson Foods, Nestle, Unilever, JBS, Cargill, Danone, Maple Leaf Foods, and Perdue, have begun to introduce their own lines of alternative protein products or entered the industry via M&As. As in the case of Beyond Meat and Impossible Foods, which followed the lead of the non-dairy milk industry, these new product lines will be positioned side-by-side with conventional meat products in supermarkets – 'merchandising in the meat case'.

Tyson Foods has launched their so-called 'blended products' line, *Raised and Rooted*, that incorporates both animal and plant-based ingredients, including a burger made from beef and pea protein isolate, and a blended chicken nugget, targeted at meat lovers and the expanding market of 'flexitarians'. One of its rivals, Perdue Farms, has also adopted this blended products strategy, sourcing some plant-based ingredients from the Better Meat Company (Lamb, 2019). This intermediate step is perhaps taken in recognition that cultured or 'clean' meat technology is unlikely to reach commercial scale in the near future. However, just to be on the safe side, meat processors Tyson Foods and Cargill have both invested in Memphis Meats, now called Upside Foods, a leading 'clean meat' start-up.

Several Big Food corporations have recently acquired plant-based protein start-ups to secure ingredients for their established brands and new product lines. Thus in 2018, Unilever bought the Dutch-owned company, Vegetarian Butcher, and is launching new products under this label – reinforced by sales of existing dairy-free brands of ice cream and mayonnaise, such as Ben & Jerry's and Hellmanns – as part of its 'Future Foods' initiative. Nestle, through its subsidiary, Sweet Earth Foods, is marketing its plant-based Awesome Burger in the US and the EU. In partnership with the Israeli cultured meat start-up, Future Meat, it is also introducing a new product line that *blends* cell cultured meat with Nestle's own plant-based meat technology (Ellis, 2021b). Another global actor, Cargill, is taking an 'inclusive' approach to the protein market by continuing to sell animal protein while introducing alternative plant-based protein products. One final example, JBS Foods, the world's leading meat processor, is marketing its Ozo range of plant-based burgers and other products nationally through Planterra, a wholly-owned US subsidiary.[14] In 2021, as a further step towards building what a spokesperson called "a global plant-based platform", JBS acquired Vivera, a leading plant-based company in European markets, for US$400 million (Askew, 2021a).

As this intense activity demonstrates, major players in the global food industry recognise that plant-based substitutes are transcending their early niche status and becoming firmly mainstream. In the words of a Nestle USA representative, "One of Nestle's strategic priorities is to build out our portfolio of vegetarian and flexitarian choices in line with modern food trends". In more direct terms, investments by major incumbents are to "*ensure that they are among the disruptors, not the disrupted*" (Stephens et al, 2019: 7; my emphasis). Big Food corporations thus are strategically poised to take a dominant role as the new 'protein economy' evolves. The plant-based meat protein market expanded by 47 per cent in 2020 to US$1.4 billion, providing further evidence of these rising trends.[15]

This emergent industry offers Big Food firms opportunities for diversification, creating production options and market presence in case consumer tastes move decisively away from animal protein, whether for health, ethical, or environmental reasons. In this context, we have seen that Tyson Foods and other leading meat processors are re-branding themselves as 'protein companies'. For example, after acquiring Lightlife Foods, which has a one-third share of the US market for refrigerated plant proteins, a representative of Maple Leaf Foods, a Canadian company known primarily for raising and processing pigs, stated that "We see it as no different than chicken or pork. *We see ourselves as a protein company first*" (cited by Burwood-Taylor, 2017; my emphasis).

The bankruptcies in 2019–2020 of two major dairy corporations, Dean Foods and Borden, provided a stark warning to Big Food since these are widely attributed to their failure to respond to the rapidly expanding sales of non-dairy milk products, which are now equivalent to 14 per cent of the conventional dairy market (Byington, 2020). The meteoric rise over the past decade of the previously little-known Swedish oat-milk producer, Oatly, is a case in point, and it is now a global brand, having entered the Chinese market in 2018 in partnership with Starbucks China and the Alibaba Group. In 2020, it sold 10 per cent of its equity to investors, including entertainment celebrities and 'influencers', such as Jay Z and Oprah Winfrey, to finance a global network of factories (Wood, 2020).[16] Oatly launched its IPO on the New York stock exchange on 20 May, 2021, significantly exceeding its anticipated capitalisation of US$9.5 billion to reach a valuation of US$13 billion in first day trading.

With the fate of Dean Foods and Borden in mind, large-scale dairy corporations and major food companies, including Arla, Lactalis, Chobani, Unilever, and General Mills, are now expanding and diversifying their existing plant-based dairy product lines. The US cooperative, Dairy Farmers of America, for example, is marketing a blended drink that combines dairy and plant-based ingredients, much like the blended brands introduced by Tyson, Perdue, and other animal-based meat companies (Coyne, 2020c). As in the case of Danone, the leading dairy companies are also active players in the M&A field.

While the pioneers of plant-based proteins, Beyond Meat and Impossible Foods, have taken significant steps to consolidate their position in US retail markets, it is doubtful that more recent start-ups can resist the predatory financial power of 'old guard' conventional meat processors and Big Food actors as they continue their rise to structural dominance over the extended protein industry (see also Howard et al, 2021).

A significant but neglected dimension of this re-structuring is that both alternative protein start-ups and established actors have patented their innovative technologies and production processes (Fassler, 2018). With 'protein conglomerates' like Tyson Foods, JBS Foods and Cargill and patent-holding start-ups, the wider protein industry is on course to reproduce the concentrated structure both of its conventional counterpart and the US food system at large (Hendrickson et al, 2019, 2020; Howard, 2021). On this trajectory, a significant segment of the protein supply chain will be privately-owned and enmeshed in restrictive, opaque intellectual property provisions, replicating conditions found in plant biotechnology and PA, for example. A litigious future is suggested by a test case of intellectual property protection in the plant-based meat sector, with Impossible Foods taking legal action in March, 2022 against Motif FoodWorks for allegedly infringing its production patent on heme, which it claims is central in reproducing the taste, aroma, and mouthfeel of animal-derived protein (Ellis, 2022a).

'Promissory narratives': disruption and paradigm change?

The AP industry is still in a formative phase – funding for a combination of 430 start-ups and established firms, such as Impossible Foods, approached US$4 billion in 2021 (Marston, 2022). Yet this emerging industry is widely anticipated to mitigate a range of environmental and human health issues caused by the high and rising production and consumption of animal protein. One view is that these benefits and attendant resource efficiencies will emerge from "a protein disruption driven by economics" that will severely disrupt, if not bankrupt, the farmed livestock sector and its value chains by 2030–2035 as the cost of alternative proteins is projected to be several orders of magnitude cheaper, reaching a factor of 10 by 2035 (Tubb and Seba, 2019: 6). However, as the authors recognise, their sector analyses and projections of price parity and beyond depend overwhelmingly on their cost curves, which are "based on limited data" and should be regarded as a "'first pass'" (Tubb and Seba, 2019: 4). At the time of writing, price parity with farmed animal protein remains a medium-term aspiration for plant-based meat analogues and a far more distant, perhaps unattainable, goal for cellular meat products.

Nevertheless, despite the problematic issue of price parity and hypothetical dietary shifts, a stream of recent reports confidently suggests that a 'protein transition' is underway, with one predicting that "it is likely that alternative proteins will capture 11% of the global protein market by 2035" (Boston Consulting Group and Blue Horizon Corporation, cited by Carrington, 2021). If this target is reached, "the report estimates that 1 billion tons of carbon emissions will have been avoided (and) farmland equivalent to the area of the UK will have been freed from supporting livestock" (Carrington, 2021). The nature and extent of this so-called 'double carbon dividend' have stimulated new debates on the benefits claimed for APs, as we discuss later.

These discursive claims represent what Alex Sexton and her colleagues call "promissory narratives" (Sexton, 2014, 2018; Sexton et al, 2019). Typically framed in strongly dualistic terms, these discourses compete with the conventional livestock industry for the ontological privilege to define "potential futures" in the agro-food system.[17] In this contested space, the benefits of a 'sustainability transition' propelled by the production and consumption of alternative proteins are often conveyed by overblown, 'iconic' abstractions. For example, the claim that the rise of APs will enhance the ability to feed the '9 billion by 2050' (Sexton et al, 2019).[18]

As Sexton et al (2019) observe, this discursive appeal to techno-science and market 'fixes' to resolve socio-ecological problems resonates with the venture capital world that nurtures these innovative firms and the popular ideology of Silicon Valley, which champions technology as "a disruptive force for societal good" (Sexton et al, 2019: 59). In these ontological politics, a 'better food system' and 'greener' world are only a technological and dietary change away. Nevertheless, putting discursive politics aside and looking at the broader picture of impending catastrophic climate change, an order of magnitude shift in protein consumption away from industrial livestock production to plant- and cell-based alternatives, notably in the wealthiest countries, would significantly mitigate global warming and the gathering intensity of its many impacts, as urged repeatedly and insistently in international fora (Swinburn et al, 2019; IPCC, 2020, 2021, 2022).

These broader patterns depend on the continued expansion of the AP industry, particularly the plant-based sector. Growth rates reached quite extraordinary levels in 2020, the initial year of COVID-19, exceeding 45 per cent in the US market, but the sharpness of the subsequent decline since 2021 has generated considerable anxiety about future prospects. In this changing context, the idealistic, visionary claims of "promissory narratives" are displaced by hard financial criteria in raising capital.

These discourses, in other words, are paramount in the initial phases of growth when start-up firms make their sales pitch to venture capital funds. Integration in this speculative world typically extends over several rounds of private funding, before firms make the transition into public capital markets by launching an IPO.[19] With this step, attention shifts from promissory discourses to the close scrutiny of corporate performance criteria – sales revenues, earnings, profits, market share – available to prospective investors in the quarterly and annual reports publicly quoted companies are required to publish.

In the current period of unease, Beyond Meat has become something of a bellwether in analyses of the slower growth in the market for plant-based meat. Briefly, the firm registered a net loss of US$181 million in the 2021 financial year, accompanied by a 78 per cent decline in its share price (Ellis, 2022b). In commenting on the 2021 earnings report, financial analysts suggested that plant-based meat firms need to improve the taste and texture of their products to convince habitual meat eaters to become permanent customers, diversify their product lines, overcome problems of scaling, and make faster progress toward price parity with animal protein (Poinski, 2022a, 2022b). Reflecting the dramatic fall in sales and earnings, Maple Leaf Foods announced that it is re-assessing future investment in plant-based food. As the President and CEO of Beyond Meat, Ethan Brown, commented, "The key question is whether this reduced growth is an aberration or a harbinger of things to come" (cited by Poinski, 2022a).

Beyond Meat's disappointing performance continued in the first quarter of 2022, with a net loss of US$100.5 million, while the US retail market for plant-based meat products grew at an annual rate of 3.7 per cent, hardly suggestive of transformative dietary change (Poinski, 2022b; Watson, 2022). In the third quarter of 2022, renewed losses and lower revenue brought layoffs of both production workers and executives (Poinski, 2022e). To add to these woes, Don Lee Farms, also a producer of plant-based protein, has launched a federal lawsuit against Beyond Meat alleging that it has gained unfair competitive advantage by making false claims about the protein content of its products (Poinski, 2022c). Such litigation tarnishes the reputation of the industry and undermines consumer confidence in plant-based meat analogues.

These rising concerns have coincided with several more realistic assessments of the promissory credentials of the AP industry. For example, the Good Food Institute (Toya et al, 2022) estimates that to achieve a relatively modest increase in the share of plant-based proteins in the global protein market from the present 1 per cent to 6 per cent by 2030 will require capital investment of US$27 billion. The report projects that 800 extrusion factories, each producing 30,000 metric tons of extruded protein annually, will be needed to reach the annual target of 25 million metric tons. In this scenario, plant-based proteins are assumed to grow at an annual rate of 18 per cent, which gives perspective to the disappointing performance of Beyond Meat and Maple Leaf Foods in 2021–2022. The GFI also warns that the industry may fail to reach this 2030 target unless it takes anticipatory action to avoid supply chain bottlenecks for key ingredients, such as coconut oil, enriched pea protein, and soy protein concentrate.

A report by IPES-Food (2022a) argues that the question of how to reduce the GHG emissions of industrial livestock production is framed simplistically as a stark choice between farmed animal protein and alternative protein. A more nuanced approach would acknowledge the complex and varied role of livestock in different production systems and geo-economic regions – for example, as a source of livelihood, a buffer against harvest failure, and an investment asset in the poorer countries of the Global South (see also Scoones, 2022).[20] The current debate takes industrial cattle production as a proxy for all farmed protein and so ignores the economic and cultural importance of other animals – chickens, goats, and rabbits, for example – in more diversified agro-food systems. This report also challenges the implicit assumption of a global protein gap in human nutrition, arguing that this frames the debate too narrowly towards corporate discourses advocating a protein transition rather than towards a systemic *sustainability* transition based on the expansion of agro-ecological production and regional and local food systems (Howard, 2022).

In analysing the resource efficiencies of alternative proteins, a key problem is the lack of hard evidence on land and water use and reduced GHG emissions. For example, little is known about the feedstock conversion ratios of AP products and the environmental and health consequences of reliance on carbon-intensive, GM feedstocks of industrial agriculture, such as soya and corn. Reviews of the few life cycle assessments (LCAs) that have been attempted under these constraints indicate that plant-based substitutes hold a substantial advantage over cultured meat, and even more so over farmed fish and farmed animal protein in terms of GHG emissions (Santo et al, 2020).[21] These authors also indicate that both plant-based and cell-based meat products would reduce their GHG footprints significantly by using sustainable sources of energy. In the case of cultured meat, Stephens et al (2018: 158) observe that "The overall picture is that (it) could have less environmental impact than beef, and possibly pork, but more than chicken and plant-based proteins".

More generally, there are strong doubts that tissue-cultured meat products can reach commercial scale. Pat Brown, CEO of Impossible Foods, rejects this possibility outright, arguing that cultured meat "will never be a commercial endeavour. The reason has to do with the fact that it is irreversibly expensive" (cited by Albrecht, 2020). This scepticism is supported by recent techno-economic analyses of the science of cultured meat and the formidable obstacles against scaling up to levels "where it can be produced cheaply enough to displace food" (Fassler, 2021). This author draws on a consultancy report by David Humbird (2021), commissioned by Open Philanthropy, which indicates that bioreactor design principles create barriers to bulk cell growth in culture and, reinforced by high capital and operating costs, lead to *"production economics that would likely preclude their affordability as food"* (Humbird, 2021; my emphasis).

These findings prompt Fassler (2021) to ask rhetorically, "Will it ever make sense to produce food the way we currently make our drugs?" On this view, "the question isn't whether companies can produce animal protein in the lab – drug companies have been doing this for decades. It is whether their approach can actually feed a meaningful number of people" (Fassler, 2021). Stephens et al (2018: 163) continue in this vein, effectively dismissing its promissory claims: "The most ambitious production target – producing cultured meat on a scale that would make a marked impact on global climate change – is likely to take many decades, if it is at all possible".

This discussion reviews various tensions that have emerged as the growth performance of the AP industry faltered in the course of 2021–2022, buttressed by calls for a more pragmatic approach to its extravagant promissory claims. In particular, this pragmatism provides an essential counterpoint to the simplistic, ideological Silicon Valley discourse that reduces the challenge of complex, multi-level global problems to dietary shifts promoted by techno-scientific innovation in protein production.

'Substitutionism 4.0': analysis and conclusion

The plant-based and cultured meat sectors of the AP industry continue the long historical movement to assert industrial control over the primary sources of food ingredients, reproducing well-established trajectories in food science and manufacture, a dynamic we conceptualised as Substitutionism (Goodman et al, 1987). From the 1970s, for example, food engineers developed sophisticated separation and product fractionation techniques that reduced food to its molecular components to provide the 'building blocks' for re-assembly into fabricated and functional foods with highly controlled nutritional constituents. Reviewing progress in protein technology at the time, Dunhill (1981: 216) reports that "potentially high-grade protein components of poor quality foods are isolated and subsequently reformed" into highly processed foods.[22]

Almost 40 years later, Spackman (2019) suggests that the AP industry is defined by 'molecular reductionism', which enables the production of novel synthetic assemblages, the epitome of highly processed, industrial foods. For this reason, together with the use of bulking agents, emulsifiers, and other additives, plant-based burgers have been labelled by some as 'vegan junk food' (Sugar, 2018; Leclerc, 2019, for example). In more temperate tones, nutritionists observe that "many plant-based proteins have high sodium levels and the bioavailability of some nutrients is questionable. Moreover, most plant-based products are highly processed, a term that consumers perceive as unhealthy" (Karmaus and Jones, 2020: 3).[23]

These comments reveal the contemporary AP industry as a quintessential expression of Substitutionism, since organisms are incorporated in assemblages that *redefine* the nature of industrial dependence on agricultural materials and so create new avenues of capitalist accumulation (Goodman et al, 1987). Alternative protein technologies relax the biological space-time constraints of the conventional animal protein economy, substituting shorter production cycles and potentially releasing agricultural space devoted to livestock grazing, feed, and fodder, relieving pressure on other natural resources.

Alternative protein production is essentially placeless, as readily urban as rural in provenance, and could re-configure the urban/rural interface dramatically. If it achieves significant market penetration, networks of brewery-like fermentation plants are likely to emerge on the peri-urban fringes of metropolitan areas and large cities.

The AP industry also exemplifies a second, recurring theme in critical agro-food studies, provoked initially by the advent of biotechnologies: the idea of nature as technics, with biophysical systems performing as technological systems in the production of commodities.[24] Boyd (2001: 632) applies this idea to the US broiler chicken industry and analyses "how science and technology have subordinated its biology to the dictates of industrial production". The AP industry would go one better by, in effect, substituting and transposing the (molecular) biology of farmed livestock from its 'parent' animal kingdom to commodity production in the plant and fungi kingdoms, leaving the conventional livestock-feed complex to adjust as best it can.

Finally, accepting both his hyperbole and implicit assumption that cell cultured meat products will be competitive, Fassler's (2018) misgivings about the trajectory of the extended protein economy deserve much closer attention from both regulators and citizens at large. Reflecting on his visit to EAT JUST in San Francisco, he observes:

> Think of it this way, a cow is the ultimate open source animal. Anyone who can afford to can raise one and sell its meat (provided they have the proper paperwork). *But the alt-protein in both the plant-based and lab-cultivated versions is intensely proprietary* … This should give us pause, heralding the rise of a new class of corporate titans, where the protein we need is not just food but intellectual property, the domain of corporations with R&D millions. What we're seeing here is, in

part, an effort to privatise what has been a public good for the entire history of our species: animals. [My emphasis]

Food-as-software describes the model of alternative proteins and processed food production more generally but its shadow of food-as-intellectual property reveals the self-serving nature of venture capital's acclamation of technological 'disruption' and its myopic perception of societal good.

5

Agri-Biotechnology and the Failed Promises of the Seed-Chemical Complex, CRISPR and Gene Editing, and Regulatory Capture

Introduction

The innovations set in train by the digital-molecular convergence are often lauded as 'disruptive' in AgTech Space and elite policy circles, such as the World Economic Forum (WEF). However, as we emphasised in Chapters 2 and 3, inter-capitalist competition in the seed-chemical complex has produced a remarkable degree of corporate continuity, with 'disruption' taking the form of successive waves of consolidation that have reduced a diverse sector to a mere handful of mega-firms in 2016–2018. This is not to under-estimate the rapid innovation in these upstream sectors, nor the inflow of venture capital that has created a fertile 'eco-system' of start-ups and research alliances. Nevertheless, as the nascent agri-biotechnology industry demonstrated so clearly in the 1980s and 1990s, the commercial future of these young innovative firms is often precarious, lacking the financial resources needed to scale-up production and expertise in logistics and marketing, making them vulnerable to the M&A strategies of large-scale corporations (see also Kenney, 1998; Howard, 2016).

This chapter draws together several common themes that run through this book. These include the mechanisms of inter-capitalist competition, manifest here in the oligopolistic domination of the global seed-chemical complex by an elite group of 'legacy' corporations, the accompanying privatisation of agricultural R&D, and the institutional capture of biotechnology governance, as detailed later. These developments buttress the industrial agro-food system, reproducing its path dependent techno-scientific trajectories and exacerbating the structural crises of rural economy and society analysed by the IATP (2020), John MacDonald and his colleagues (2017, 2018, 2020), Jane Gibson (2019), and Philip Howard (2021).

Here, we again meet the likes of Bayer-Monsanto and the Syngenta Group, with their siren promises for successive generations of plant biotechnologies, locking-in farmers on the treadmill of competitive innovation. Finally, in the form of gene driving, we encounter a Janus-faced technology that offers *commercial* continuity for plant protection chemicals but also heralds a radical *ecological* discontinuity. Gene driving transforms the spatiality of molecular intervention from individual plant and animal species to engineering the extinction of insect-vectored plant pests at the level of agro-ecosystems, and beyond. This powerful assemblage, secured by intellectual property rights, gives fresh meaning to the old charge against biotechnology of 'playing God'.

Vertical integration and concentration in the 'seed-chemical complex'

The antecedents of this now highly consolidated complex are to be found in the rapid diffusion of the 'genetic-chemical' technology of hybrid seed corn in the American Mid-West farm belt in the late 1930s and 1940s. This innovation, with the 'biological lock' of hybrid vigour, prompted an early round of consolidation in the seed industry and subsequently led to the transfer of near-market R&D from public agricultural research stations to the private sector (Kloppenburg Jr., 1984, 1988; Goodman et al, 1987). When the major agro-chemical and pharmaceutical corporations began to integrate vertically into the emerging agricultural biotechnology industry in the 1980s, leading seed companies were targeted for their elite germplasm and marketing networks, while the acquisition of start-ups spun off from university research labs gave access to new plant traits.

The diffusion of genome sequencing techniques from the mid-1990s and gene editing after 2010 accentuated this capital-intensive life sciences trajectory, as companies sought both to defend their existing assets and market share and capture new value streams in this changing 'genes-to-seed' space. Vertical integration has gained further momentum from the farm service platforms developed by these same actors in response to farm-level digitalisation and its impacts on markets for seeds and agro-chemicals, as we discussed in Chapters 2 and 3.

The industrial context of agricultural biotechnological innovation over the last 40 or so years is represented by the consolidation of the 'seed-chemical complex' as a powerful, vertically integrated, multi-faceted actor, and the *parallel* privatisation of molecular biology and its commercial applications. This is manifest in the formidable legal infrastructure established since 1980,[1] which is directly influenced by the 'informationist', 'life as code' approach to molecular genetics.[2] Life science corporations have erected intellectual property platforms holding patent rights over novel genetically engineered life forms and DNA sequences,[3] complemented by portfolios of elite germplasm, a revenue-generating licensing system of transgenic crop traits, and material transfer agreements[4] to prevent farmers from saving GM seed.

These intellectual property platforms are the cornerstone of the biotechnology industry and its commercial *modus operandi* in the innovation process. Life science corporations now have large-scale in-house R&D facilities that collectively generate a constant and substantial flow of patent applications and cross-licensing agreements. A second pathway exemplifies the continuing significance of universities in molecular biological research and the permeable boundaries between public and private R&D, designated by Martin Kenney (1986) as the 'university-industrial complex' in biotechnology. This pathway might begin with a scientific breakthrough in a university lab, followed by a university patent application, the subsequent concession of licencing rights to the commercial start-up or 'surrogate' firm established by the innovative university scientists, and licencing agreements with major life science corporations.

In the case of gene editing and clustered regularly interspersed short palindromic repeat (CRISPR) technology, these steps take us from the lab of Dr Jennifer Doudna at the University of California, Berkeley, and her co-inventor, Dr Emmanuelle Charpentier of Vienna University, to the hard-fought and continuing legal disputes between the University of California and the Harvard/MIT Broad Institute over ownership of the foundational patent (Cohen, 2020; Sherkow, 2022). Meanwhile, UC Berkeley has granted licencing rights to Dr Doudna's start-up, Caribou Biosciences, which has reached licencing agreements with DuPont /Pioneer, now Corteva Agriscience, and other life science companies. It appears that the oligopolistic life science corporations already control the patent and licensing landscape of gene editing applications in agriculture (Egelie et al, 2016), which "are hewing to familiar patterns of the past fifty years" (Montenegro, 2020b).

We take a closer look at this political economy in the following pages, beginning with the commercial introduction of the 'first generation' of plant biotechnologies and the emergence of the so-called 'new breeding technologies', notably gene editing systems, such as CRISPR-Cas9, and concluding with gene drives and the corporate capture of the US regulatory system. As we will find, these recent innovations have again exposed the pronounced differences in the regulatory and governance cultures of the EU and the US. We explore these issues and the departing Trump Administration's reforms of regulatory oversight in biotechnology in the final section of this chapter.

'First generation' plant biotechnologies

The promotional tropes of plant biotechnology in the 1990s typically compared the new transgenic techniques with conventional plant breeding, claiming that science could now harness the 'power of nature' to 'feed the world', enhance environmental sustainability and introduce new crop traits to mitigate abiotic stress in resource-poor environments, launching a second 'green revolution'. However, more pragmatically, biotechnology offered the seed-chemical corporations the commercial advantages of greater speed and precision in plant breeding, and these remain key economic incentives 20 to 30 years later with gene editing technologies.

Greater speed in plant breeding meant lower costs and shorter time lags between lab research, test plots and the marketing of commercially valuable transgenic crop traits. "Agricultural biotechnology, in short, held out the potential for both *accelerated product cycles and new value-added products*" (Boyd, 2003: 33; my emphasis). The advantage of faster R&D processes was closely allied to more precise methods of developing new crop cultivars. Biotechnology made it possible to select DNA from one organism to be incorporated and expressed in the DNA of a different organism rather than being restricted to the random recombination of DNA achieved by crossing compatible parent lines. As Kloppenburg Jr. (2004: 192) comments on this period, "New plant varieties are being engineered in the strongest sense of that word's connotations of precision and foresight."

However, since life science companies are in the business of selling seeds and pesticides, the heart of the promotional message was addressed to farmers: innovative crop protection technologies – glyphosate-tolerant and Bt-resistant[5] plants – promised to raise farm incomes by increasing yields and reducing pesticide costs. In fact, these transgenic crop protection traits for major row crops proved to be the principal, and commercially by far the most successful, innovation of first generation plant biotechnologies, and were quickly disseminated globally, with the major exception of the EU. It is the impacts on environmental and human health of these herbicide-tolerant and insecticide-resistant traits and their agro-chemical 'packages' that have attracted such intense opposition, as we will discuss later.

Failed promises

How did farmers fare with transgenic pest management systems? To address this question, we draw on contemporary assessments made during the first 10 to 15 years following their commercial introduction in 1996.

Farmers' operational decisions are driven by comparisons between available pest management systems, and a number of studies emphasised the advantages of adopting the simplified, GE seeds-based, pest control technologies in the later 1990s. For the period 1996–2013, Fernandez-Cornejo et al (2014) found that herbicide-tolerant and Bt insect-resistant varieties, later reinforced by GM seeds with multiple ('stacked') traits, had protected yields and raised net returns when compared to planting conventional seeds. Examining the efficacy of GE pest management systems in the years 1996–2005, Benbrook (2018: 387) reports that they provided farmers with new "options that were extremely effective, easy to deploy, robust and roughly the same cost as alternative methods of dealing with the same weed and insect pests". He adds that the GE crop protection technologies reduced reliance on "almost assuredly more toxic alternatives" before the emergence of resistant target pests eroded their effectiveness.

With these initial advantages, adoption rates were rapid and unprecedented. By 2013, herbicide-tolerant varieties were planted on 93 per cent of the soybean acreage in the US, 85 per cent of the corn acreage, and 82 per cent of the acreage devoted to cotton (Fernandez-Cornejo et al, 2014). Insect-resistant (Bt) cotton accounted for 75 per cent of US cotton acreage and Bt corn for 76 per cent of corn acreage (Fernandez-Cornejo et al, 2014). In short, US farmers of staple row crops became ever more firmly locked into the mono-cultural agro-industrial model, with their livelihoods reliant on the efficacy of bio-chemical pest resistant technology, extending the power and control of the oligopolistic seed-chemical complex. In 2011, Monsanto, DuPont/Pioneer and Syngenta controlled over half of the global market for commercial seeds, and with Dow, BASF, and Bayer, accounted for 75 per cent of global agrochemical sales (ETC Group, 2013).

Yet the early promise of GE-based seed protection technologies was soon squandered by the build-up and spatial spread of glyphosate-tolerant weeds – 'super weeds' – and Bt-resistant insect pests. For Benbrook (2018: 390), this outcome was a matter of regulatory omission and corporate lobbying as the glyphosate and Bt-resistance monitoring strategies[6] that had been effective in the period 1996–2005 were abandoned, setting "the stage for the collapse in first-generation GE-herbicide resistant technology".[7]

With their patent control of the genetics of the staple row crops, seed-chemical firms responded with a further iteration of first generation GM technology by introducing seeds 'stacked' with multiple resistance traits, including crops that tolerate the highly toxic phenoxy herbicides, dicamber and 2,4-D, leaving farmers with little choice but to bear the rising costs of staying on this accelerating technological treadmill (Cochrane, 1979; Benbrook, 2018: 390). "*Overall, pesticide use has about doubled, led by the huge increase in the volume of glyphosate applied*, coupled with the growing need for two to four or more additional herbicides to deal with growing infestations of glyphosate-resistant weeds" (Benbrook, 2018: 390; my emphasis. See also Waltz, 2010; Benbrook, 2012; Harker et al, 2012). GM corn with stacked traits grew from 1 per cent of US corn acreage in 2000 to 9 per cent in 2005 and 71 per cent in 2013 (Fernandez-Cornejo et al, 2014: 7).[8] Such herbicide 'cocktails' of multiple active ingredients, when used as a formulated product, carry unknown health, ecological, and environmental risks.

This outcome, with farmers trapped on the treadmill of GM crops, can be set in a wider context by drawing on Heinemann et al (2014), who argue that the choice of GM-biotechnology packages has not benefitted US farmers when compared with their Western European counterparts.[9] Reviewing the parameters of yield, pesticide use, and germplasm diversity for the period 1961–2010, the central finding is that "The US (and Canadian) yields are falling behind economically and technologically equivalent agro-ecosystems matched for latitude, season and crop type; pesticide use (both herbicide and insecticide) is higher in the US than in comparator Western European countries" (Heinemann et al, 2014: 84).[10] This study also finds evidence of stagnation and decline in germplasm diversity, which the authors attribute not to GM crops but rather to the transition from public sector breeding programmes and a robust seed-saving and exchange culture to private breeding and the restrictive effects on innovation of intellectual property platforms.

In sum, innovation policy in US staple row crops increasingly has been devolved to the private sector since 1980 and strongly orientated to the demands of high external input, uniform agro-ecosystems (Heinemann, 2020). Correspondingly, intellectual property rights and other patent-like protections, including the 'no-replant' clause enforced by Monsanto and its corporate counterparts, have been strengthened, leaving farmers dependent on the GM-seeds-chemicals industry and reducing their control of the farm production process (Mascarenhas and Busch, 2006).

Focusing on a second plank of US agricultural policy, since GM coverage has reached "near saturation" levels for staple commodity crops, public subsidies essentially are underwriting GM crop production in a mutually reinforcing cycle (Key and Roberts, 2007; Heinemann et al, 2014). Taking the distribution of the top 20 main subsidy programmes by farm size, 86 per cent of this funding is received by the largest 20 per cent of recipients (Howard, 2016: 9). From a global viewpoint, public subsidies for the production of animal feed, such as corn and soya, benefit the intensive grains-livestock economy and thus contribute to anthropogenic GHG emissions.

This analysis sets the stage for the following discussion of gene editing, aka the "new breeding technologies", as US crop management systems enter the second stage of engagement with GE seed-based technology. With US farmers caught in a cost-price squeeze, the stakes are high: "If the pest resistant management strategies fail in the second generation GE crops, as they failed for the first generation, the technological treadmill will turn more rapidly and at greater cost to farmers" (IATP, 2020: 3–4). Again, this course has effectively been rejected by the EU, as we see in the final section on issues of regulation and governance.

Genomics, gene editing, and gene driving

The lobbying organisations funded by the biotechnology industry again paint a glowing picture of the future that awaits once gene editing technologies become widely adopted. This 'silver bullet' status is heightened by the admission that first generation genetic engineering has failed to live up to its earlier billing and is now openly disparaged for its clumsiness. "Genetic engineering is typically ham-fisted: it often involves inserting a section of DNA from an entirely different kind of organism ... with little control over where in the genome it lands." Compared to genetic engineering and conventional plant breeding, "genome editing offers both *subtlety and speed*, wherever on the genome the researcher wishes to target" (Ainsworth, 2015: 15; my emphasis).

However, as Steinbrecher (2015) observes, precision is not equivalent to predictability and safety when gene-edited crops are released into the open environment. Contrary to industry claims, these new techniques do involve risk, uncertainties and unintended effects on health and the environment that warrant regulation.[11] They also open a Pandora's Box of gene drives and 'ecological engineering', as we will see later.

We have alluded to the rise of genomics as the transformative force in the biological sciences and innovation since the early 2000s: "[The] ability to determine an organism's complete genetic makeup has changed the way science is done" (Archibald, 2018: xv). This author describes genomics as a rapidly evolving set of research methods, DNA sequencing technologies and instrumentation, where dramatic increases in computing power and data storage have reduced the time and costs of sequencing. "The genomes of simple bacteria and viruses can be sequenced in a matter of hours on a device that fits in the palm of your hand" (Archibald, 2018: xv).[12,13] Speed and precision again are highlighted as the watchwords of techno-scientific innovation in biotechnology (see also Nasti and Voytas, 2021).

High throughput DNA sequencing and related ICT innovations have generated progressively larger and more diverse sets of functional genomic data. This rapidly accumulating wealth of genetic information is the raw material for gene editing techniques, such as CRISPR–Cas9. As one university research scientist observes, "It really opens up the genome technology of virtually every organism that's been sequenced to be edited or engineered" (cited by Dance, 2015: 6245).

Gene editing extends the scale and scope of biotechnology beyond conventional genetic modification, and unleashes commercial pressures to undertake so-called 'ecosystem engineering' of farm landscapes (ETC

Group, 2018a). CRISPR-mediated gene editing tools *amplify* the technical capabilities of synthetic biology and gene driving by facilitating the insertion of engineered DNA *constructs* into genomes, opening up wider, more diverse applications in crop and livestock production. For Pixley et al (2019: 172), gene editing is a revolutionary enabling step that allows the full potential of genome sequencing to be realised by integrating *reading* DNA with DNA synthesis and *rewriting* DNA into genomes "on unprecedented scales". In pest management, for example, these authors look forward to "setting up viral roadblocks to *multiple* viruses instead of the *one-to-one* resistance strategy that is currently common" (Pixley et al, 2019: 173; my emphasis).

Several gene editing techniques, including ZFN and TALEN, were in use before the advent of CRISPR.[14] These earlier technologies are expensive as a new ZFN or a TALEN protein must be produced for each edit, a difficult and error-prone process. CRISPR is incomparably simpler and more versatile, and it is far cheaper to create an ribonucleic acid (RNA) sequence than a ZFN or TALEN protein. CRISPR technology involves construction of an RNA template that matches the target DNA sequence in the genome of the cell, and these matching sequences can then bind to each other. The RNA segment of the CRISPR, or 'guide RNA', directs the Cas9 enzyme (CRISPR-associated protein) to the targeted DNA sequence to cut the organism's genome at this location and make the edit (NHGRI, 2017). By controlling how this break or incision in the 'edited' DNA is repaired by the cell, CRISPR can be used to remove or alter genetic material and to insert new engineered DNA sequences into the genome. As Montenegro (2019: 1059) observes, "With the discovery of the guided Cas9 enzyme, researchers have found a gene editing tool that works across numerous kingdoms, from plants and fungi to insects and animals."

Given the past concentration on crop input traits and their contribution to sales revenues, perhaps it is not so surprising to find that the first commercial gene-edited plant is a herbicide-resistant oilseed rape. Further crop innovation along this well-trodden path is likely since, as Derek Jantz, co-founder of Precision BioScience, notes, "All plants have an analogue of the EPSPS gene that is inserted into Monsanto's Roundup Ready crops. It should be possible to create similar herbicide resistance by editing a plant's own version rather than bringing in an external gene" (cited by Cressey, 2013).

Herbicide-resistant oilseed rape was introduced commercially in 2016 by a Californian company, Cibus, using proprietary gene editing technologies, RTDS™: (Rapid Trait Development System), which, according to its website, "develops nature-identical plant traits that are indistinguishable from those that would occur in nature". Just to emphasise the IP defences erected by Cibus, this website observes that "Our technologies are protected by over 300 patents and patent applications across 16 patent families."

Following USDA regulatory approval in 2016, the gene edited oilseed rape is marketed as a non-GM crop on the grounds that the resistance trait results from mutagenesis, without the insertion of a gene from a foreign organism. This distinction is at the crux of current regulatory differences between the US and the EU.

However, a major credibility gap persists at the heart of these narratives that has yet to be bridged: the promise that genome editing technologies can respond to the challenges of global warming by introducing agronomic traits that enhance plant resistance to drought, salinity, and soil acidity, and raise nutritional efficiency. To improve the resilience of cropping systems is a daunting task, however, since "multiple genetic and epigenetic mechanisms … control plant resistance to abiotic stress" (Dunwell, 2011: 1). In fact, applications of genome editing to engineering complex, polygenic traits are limited in scope, since "like first generation genetic engineering techniques, genome editing operates *outside the organism's existing regulatory framework that controls gene expression*" (Cotter et al, 2020: 20; my emphasis). For these authors, "Modern conventional breeding techniques such as genomic selection and marker-assisted selection are more suited to the breeding of complex traits" (Cotter et al, 2020: 20).[15,16]

These brief comments provide a more realistic assessment of genome editing, qualifying claims that these techniques will trigger an imminent 'techno-fix' to mitigate the damaging impacts of global warming on cropping systems, and thereby 'feed the world'. A similarly critical and cautionary perspective is also needed as agro-eco-systems are drawn inexorably into the 'brave new world' of gene driving as the momentum of scientific and industrial interest accelerates.

Gene driving and engineering agro-ecosystems

Although the notion of gene drives is not new (Burt, 2003), the ease and low cost of genome editing, and particularly CRISPR-Cas9 – the 'molecular scissors' – has focused attention on gene drive strategies to control insect-vectored human pathogens, such as malaria, dengue fever, and the Zika virus. In August, 2020, regional and state agencies in Florida and the federal US Environmental Protection Agency (EPA) approved a pilot programme to release synthetic gene drives developed by the US-owned company, Oxitec, to reduce mosquito-borne malaria in the Florida Keys (LeMieux, 2020).[17] One research scientist, Omar Akbari of the University of California, San Diego, hailed this landmark decision as "paving the yellow brick road" for engineered gene drive systems (cited by LeMieux, 2020). In turn, these developments advance the prospect that similar approaches soon will be used to suppress sexually-reproducing, insect-vectored plant pathogens and other invasive species in industrial agriculture (see also Pixley et al, 2019).

The aim of gene drives is to engineer wild populations by releasing organisms with synthetic genetic elements into invasive, sexually-reproducing populations of insects, plants, and animals so that future progeny spread the novel genetic material through succeeding generations and 'drive' the target species to extinction (Cotter et al, 2020). Synthetic gene drive systems are engineered with 'selfish' genetic constructs that do not follow Mendelian patterns of inheritance since the probability of specific genetic material being transmitted is increased above the normal 50 per cent in the progeny of organisms that reproduce sexually. In short, a gene drive is used to bias inheritance in favour of a specific heritable element so that the engineered transgene becomes more prevalent in the population over future generations (Alphey et al, 2020).

In sexually-reproducing species, the normal process of inheritance "is the cornerstone of biological diversity ... But gene drives force a species towards uniformity or extinction, ... a violation of the fundamentals of evolution" (ETC, 2016). The ethical presumption that "humans can and should use such powerful, unlimited tools to control nature ... will change the fundamental relationship between humanity and the natural world forever" (ETC, 2016: 2; see also Medina, 2018).

Several different gene drives are in development whose temporality and scale, at least in principle, can vary from 'global', where the target species is driven to extinction, effectively a form of 'terminator technology', to 'local' gene drives, which are meant to pass on hereditary traits to only a designated number of generations. Whereas 'conventional' biotechnology involved the genetic modification of individual crops and livestock species, gene drives introduce spatial and temporal dimensions of quite a different order. In other words, biotechnology can now engineer entire agro-ecosystems by intervening in the environment that surrounds domesticated farm crops and livestock (ETC Group, 2018a; Montenegro, 2019). This *new spatiality* raises the risks and potential harms to the biodiversity of agro-ecosystems on to a higher, more dangerous plane. In the words of a gene drive researcher, "Gene drives are intrinsically about altering the shared environment" (cited in Montenegro, 2019: 1065).

These concerns are echoed by Power (2021: 1; my emphasis), who observes that

> CRISPR-Cas gene editing tools have brought us *to an era of synthetic biology that will change the world* ... We do not know how genomic processes (including regulatory and epigenetic processes), evolutionary change, ecosystem interactions, and other higher order processes will affect traits, fitness, and impacts of edited organisms in nature.

In a call to arms, she suggests that scientists need "To anticipate how 'synthetic threads' will affect the web of life on Earth ... [and] confront complex system interactions across many levels of organisation" (Power, 2021: 1). In her view, the scientific community currently is ill-equipped to meet this challenge.

Research also is underway to develop so-called reversal or immunisation drives to offset unintended and undesirable effects of a previous drive but "the effectiveness of these remediation strategies is theoretical and any ecological damage … could be irreversible" (Medina, 2018: S256).[18] A report from the US National Academies of Sciences, Engineering and Medicine also is sceptical and argues that there is insufficient evidence available to support "the release of gene drives into the environment" (NASEM, 2016: S257). Others also have emphasised the complexities of conducting reliable ecological risk assessment. For one prominent gene drive scientist, "The very idea of a field trial is that it's a trial confined to an area. Our model indicates that this is not the case" (Cited by Zimmer, 2017).[19] Not surprisingly, given the lack of effective spatio-temporal control, there have been widespread calls for a global moratorium on field releases, the inclusion of rigorous biosafety protocols in national regulatory policies and international agreements, and wider, inclusive, cross-cultural governance systems.[20]

Notwithstanding, the Pandora's box of gene drives is now well and truly open. As we have seen, following the regulatory reforms introduced by the Trump Administration in 2020, the EPA has approved pilot release programmes of GM mosquitoes in the Florida Keys (Kuzma, 2019a, 2019b; 2020), Texas and, in 2022, California.[21] Moreover, agribusiness pressures are mounting to use CRISPR-mediated gene drives to re-engineer agro-ecosystems in ways that consolidate the industrial agriculture model. For example, by engineering gene drives to reverse the herbicide and pesticide resistance acquired by invasive weeds and insects, thereby extending the commercial lifespan of agrochemicals, and releasing new drives periodically to counter natural selection and field-evolved resistance.

This scenario is elaborated in a penetrating and cautionary analysis by the ETC Group and the Heinrich Boll Foundation, *Forcing the Farm* (2018a), which warns of the enormous ecological risks and harms that could arise if gene drives are widely adopted as pest management strategies in agro-ecosystems. This analysis identifies farm landscapes as the leading commercial arena of the future: "*The two foundational patents for gene drives are largely written with agricultural applications in mind*" (ETC Group, 2018a: 4; my emphasis). Each of these patent applications – by Harvard and MIT's Esvelt and the University of California's Bier. Gantz and Hedrick – enumerates literally hundreds of weed, insect, mollusc, and nematode species that could be targeted by gene drives. "The foundational patent application on RNA-guided gene drives by Esvelt lists more than 180 agricultural weed species that might be targeted by CRISPR gene drives, as well as 160 … pest species relevant to agriculture" (ETC Group, 2018a: 16).

These portents foreshadow a far-reaching strategic transition in agro-ecosystem management that complements and strengthens the main commercial axes of the 'seed-chemical complex'. Based on the potentialities designated in the two foundational patent applications, gene drives could remove or reduce the threat of pest resistance to the yields and productivity of mono-cultural commodity production. Although agribusiness may choose for commercial reasons to continue 'stacking' herbicide tolerance traits, gene drives offer the possibility of 'sterilising' crop environments, re-defining the very meaning and substance of agricultural biodiversity. Once released into agro-ecosystems, gene drives almost inevitably will create new and unknown socio-ecological realities, since they are synthetically designed to spread and persist, can mutate, and may move between farmed and natural ecosystems. Alarmingly, the risks and dangers of gene drives to the integrity of the biosphere are emerging in a "governance vacuum" (ETC Group, 2018a: 30). The journey to this point is discussed in the next section.

Regulation and governance: institutionalising corporate power

This chapter has stressed the complementarities forged between modern genetic research, imbued with the paradigmatic concept of 'life as code', and the post-1980 intellectual property rights (IPR) regime. These synergies between science, technology, law, and political economy have fostered the privatisation of biotechnology and pronounced industrial concentration. Here, we examine the institutional framework that has generated these synergies and established a commercial landscape dominated by the IPR portfolios of the 'Big Four' life science corporations: Bayer/Monsanto, the Syngenta Group, Corteva Agriscience, and BASF. It is a landscape characterised by the extraction of economic rents, oligopolistic behaviour, regulatory capture, and corporate aggrandisement.

For decades before 1980, patent protection in industrial sectors existed separately from the legislative framework of plant breeders' rights and, with some exceptions,[22] this separation was enshrined in the International Convention for the Protection of New Plant Varieties (UPOV) of 1961. This institutional divide was comprehensively dismantled in the US by a series of ground-breaking decisions in the early 1980s. These extended patent protection to genetically engineered living organisms, overturning the 'product of nature' principle, which had long dominated this field of patent law, provoking ethical critiques of 'patenting life' and 'patenting the biosphere' (Juma, 1989).

The first of these decisions arose from the landmark case of *Diamond vs. Chakrabarty* in 1980 and the narrow ruling by the US Supreme Court that "a live human-made micro-organism is patentable subject matter" (cited by Kloppenburg Jr, 1988: 262). Further clarification followed in 1985 with the decision by the US Board of Patent Appeals and Interferences in *Ex Parte Hibberd*, which extended utility patents protection to plants and allowed plant "breeders to choose among the separate statutes for the best form of protection" (Kloppenburg Jr, 1988: 263). Utility patent legislation encompasses claims not only to the plant variety but also to its constituent parts, including DNA sequences, genes, cells, and seed. As Kloppenburg Jr (1988) emphasises, the *Hibberd* decision was instrumental both in stimulating firms to establish a portfolio of patent rights as a source of licensing fees or 'rents', and enabling seed companies to enforce property rights on GM seeds, breaking the long-standing tradition of farmers saving harvested seed for replanting.

This radically new situation is neatly summarised by Boyd (2003: 40–1):

> *Chakrabarty* and *Ex Parte Hibberd* thus created a much more formal and increasingly privatised system of intellectual property protection for germplasm resources and new genetically engineered organisms. Incentive structures and revenue streams have shifted to private actors able to capitalise on the new technologies, facilitating a race to patent valuable resources and exploit first-mover advantages.

By its very nature, however, this race can be run only by well-resourced corporations, erecting formidable barriers to entry, and driving economic concentration. Two further legal developments in 1980 reinforced the legal infrastructure of this emerging industry.

The Cohen-Bayer patent, which was granted on novel organisms created by recombinant DNA technology, covered both the organisms and the techniques used to produce them, exploiting the commercial opportunities afforded by the new IPR regime.

Moreover, this patent also demonstrated that universities potentially "could enrich themselves by licensing the inventions of their scientists" if restrictions on the patenting of federally-funded research were abolished, which occurred with the passage of the Bayh-Dole Act that same year (Boyd, 2003: 41).[23] This legislation established the foundations of the 'university-industrial complex' (Kenney, 1986), complementing the *Chakrabarty* IPR regime and "reinforcing the tendency toward privatisation of biotechnology research and creating significant incentives for universities to develop their own intellectual-property portfolios and licensing strategies" (Boyd, 2003: 42).

Hollowing out governance

In an ideal world, the institutions of governance would pursue and defend the public interest and be articulated through democratic, transparent, participative, and inclusive processes. In the case of US agricultural biotechnology, however, such processes are conspicuously absent by design, and governance has been captured and moulded in the interests of dominant corporate actors. This was already becoming apparent in the 1970s, some 20 years before the commercial application of recombinant DNA technologies, when the regulatory debate was framed in technical terms of biohazards and their containment rather than questions of wider social interest and the ethics of using these powerful tools to modify the genomes of life forms (Wright, 1994). This technocratic discourse prevailed at the 1975 Symposium on Science, Ethics, and Society convened by research scientists at Asilomar in California, and has since continued to dominate US regulatory policy as corporate interests gained in strength and issues of international competitiveness became more prominent (Wright, 1994; Kevles, 1998; Kelso, 2003. See also Jasanoff and Hurlbut, 2018).

Following the Asilomar conference, which called for stringent self-governance, the National Institutes of Health (NIH) issued voluntary guidelines for R&D safety procedures, although these were strictly applicable only to recipients of NIH funding.[24] These national controls were short-lived, however, as the 1980 *Chakrabarty* decision and the prospect of commercial development prompted Congressional hearings to review the oversight mechanisms for the release of GM life forms. Concurrently, the White House and the Office of Science and Technology Policy formed an interagency working group to consider the oversight role of federal agencies, which led to the introduction in 1986 of the Coordinated Framework for Regulation of Biotechnology (Wolt and Wolf, 2018).

The Coordinated Framework radically changed the focus of regulation in favour of corporate interests, moving away from the process-based oversight approach, exemplified by the Asilomar conference proceedings and NIH guidelines, toward a product-based approach. Under the new framework, GM products would be reviewed by the USDA/APHIS,[25] the Environmental Protection Agency (EPA) and the US Food and Drug Administration (FDA). With this change toward a product-based system and the presumption of 'substantial equivalence' between GM organisms and conventionally-bred organisms, the onus of proof shifted dramatically from the biotechnology industry to the public to demonstrate that regulation is warranted for reasons of human health, ecological harms, or ethical concerns. For Kelso (2003: 242), "having eliminated most social issues from the agenda", consumers and activist organisations (NGOs) were left as "imperfect surrogates for democratic institutions" (Kelso, 2003: 239). This surrogacy found some expression in the social movements campaigning for the labelling of GM foods and Right-to-Know legislation (Guthman, 2003; see also Schurman and Monroe, 2003).[26]

The transatlantic divide in regulatory policy

Oddly enough, the legacy of Asilomar can be recognised in the EU's adoption of a precautionary, process-based, regulatory approach in which the technical method used in genetic modification is the foremost concern of risk assessment. Applications to commercialise GMOs are evaluated by the European Food Safety Authority (EFSA) and, following field trials and risk assessment, Member States vote on whether to authorise the release of the GMOs into the environment. Since 2015, national EU governments can 'opt out' of a decision to cultivate a GM crop on societal grounds, although social concerns are not formally included in the science-based remit of EFSA. This 'opt out' provision reflects the deep historical roots of opposition to agricultural biotechnology in EU member countries (Levidow et al, 2000; Reed, 2002, 2008; Peschard and Randeria, 2020) and creates political space for citizen groups and NGOs. Nevertheless, "at present, there is no mechanism for societal consultation on GMOs in the EU" (Cotter et al, 2020: 29).

However, there is more to this transatlantic divide in the politics of risk and regulation. Whereas the precautionary principle is the foundation of risk assessment in the EU, Kelso (2003: 245) emphasises the formidable difficulties confronting efforts in the US to secure regulatory oversight on a precautionary basis. "In effect, this system protects not the public but the intellectual property and commercial opportunities of the applicant." Noting that international competitiveness is a significant concern of US regulatory agencies, Kelso (2003: 245) adds: "Reinforced by this industry-centred view of regulatory oversight, the structure of the regulatory framework, the burden of proof, and the proprietary control of information all impede or nullify opportunities for the practice of democracy." *Although not framed in these terms, this conclusion is a powerful statement of regulatory capture.*

The strong regulatory differences between the US and the EU were strikingly exposed in the prolonged debates and lobbying over the regulation of genome editing, or 'new breeding techniques' (NBTs). Before genome editing, and notably CRISPR-Cas9, foreign plant genes were transferred into GM plants using either plant pests, notably *Agrobacterium tumefaciens*, or by direct methods, such as particle bombardment (Dunwell, 2014). The regulatory authority of the USDA derived from the use of *Agrobacterium* and other plant pests to transfer DNA into GM plants. However, gene editing techniques do not insert DNA from other organisms into the target plant genome but rely on mutagenesis, which is widely used in conventional plant breeding. This is the basis of claims that gene editing simply imitates the process of mutation that occurs in nature and so gene-edited plants should not be regulated as GMOs.[27] Rather, risk assessment should be based on the characteristics or phenotype of the new plant variety and the safety and environmental impact of its novel traits. As we saw earlier in the case of CIBUS, its gene edited oilseed rape plant is regulated as a non-GM crop.

As Ainsworth (2015) observes, this distinction between mutagenesis and genetic modification was at the heart of the lobbying campaign – 'Embracing Nature' – funded by the biotechnology industry and multinational life science corporations to convince the European Court of Justice (ECJ) that gene edited crops should not be regulated as GMOs (Corporate Europe Observatory, 2016, 2018). However, with strong support from NGOs in EU countries (see also ENSSER, 2017), the ECJ ruled on 25 July, 2018, that gene edited plant varieties and other organisms should be assessed for health and environmental risks in the same way as genetically engineered plants and be regulated as GMOs in accordance with existing EU practice under Directive 2001/18.

Failure to secure 'light-touch' product- or trait-based regulation for gene edited crops, as in the US, intensified the pressure on the European Commission to reverse this decision, with interventions by the US Secretary of Agriculture, Sonny Perdue, (Corporate Europe Observatory, 2019; see also Galinsky and Hillbeck, 2018) and disparaging statements from the 'Big Four'. These are exemplified by the CEO of the Syngenta Group:

> What worries me at the moment is that the EU is moving away from science and risk-based regulation of technology. Instead, it seems that they are taking agriculture backwards so that farmers are losing key tools to be able to grow crops in an environmentally sustainable way. We need to carefully look at and trust in science. (AgFunder News, 2019)[28]

De-regulation

In fact, it is the US that is now moving even further away from risk-based regulation and public oversight towards corporate self-governance. This is abundantly clear from the recent regulatory reforms introduced by Executive Order of the Trump Administration on 11 June, 2019, and finalised on 18 May, 2020. The new rule, SECURE,[29] runs counter to the proposals made by the Obama Administration in January, 2017, to retain plant-pest and noxious weed risk assessment, which the biotechnology industry had opposed as 'stifling innovation' (Kuzma, 2019a). The Trump reforms recognise that gene-editing technologies have circumvented regulatory oversight by the USDA, which was originally triggered by concern that the use of plant-pest DNA sequences would pose a pest risk to plant health (Wolt and Wolf, 2018). *With this new rule, any remaining vestiges of public oversight are effectively delegated to corporate interests.*

Not only is the thrust and language of the new rule "decidedly anti-regulation and pro-biotechnology" (Kuzma, 2019b), but the pretence of governance as public participation and transparent decision-making no longer bears scrutiny. Any such notion is lost in the slipstream of innovation and corporate power (Montenegro, 2020c). Henceforth, corporations and scientists will determine whether or not a new product qualifies for exemption from regulation, consigning USDA/APHIS to an advisory role. As Greg Jaffe of the Center for Science in the Public Interest notes, "The result is that government regulators and the public will have no idea of what products will enter the market and whether those products appropriately qualified for exemption from oversight" (cited by Skotstad, 2020). Indeed, the USDA anticipates that virtually all biotech crops will receive exemption from field testing and data-based risk assessment (Kuzma and Grieger, 2020).

The Trump reforms of the Coordinated Framework and the final evisceration of regulatory oversight occur at a particularly hazardous time as the issue of gene drives comes to the fore. We have seen that the EPA granted an experimental release permit on May 1, 2020 to the US-owned company, Oxitec, "to release millions of GM mosquitoes every week over the next two years in Florida and Texas" (Kuzma, 2020), and approved open environmental release in California in May, 2022. According to the company website, male mosquitoes are engineered with a 'self-limiting gene' that prevents the offspring of female mosquitoes from surviving to adulthood. Under the heading, "Time to reassess risk assessment", Kuzma (2020) draws attention to the lack of political accountability and writes that "The closed nature of this risk assessment process is concerning to us."

Analysis and conclusion

This critical observation brings the trajectory of biotechnology regulation and governance full circle, providing a powerful case study of regulatory capture. Almost two decades earlier, Boyd's (2003: 28) view that there is a "deep monopoly structure" to the agricultural biotechnology complex is more convincing than ever.

This structure has catalysed the "re-location of power" (Gibson, 2019: 25) over the farm production process to an ever-shrinking elite of life science corporations. As we emphasised here and in Chapter 3, the privatisation of agricultural R&D and commercial logic have combined to create formidable levels of industrial concentration in crop protection for the major Mid-Western row crops – corn, soya, canola, and cotton – effectively locking farmers into a technological cul-de-sac. PA platforms are part and parcel of this trajectory, essentially marketing devices for the seed-chemical complex.

Agricultural applications of gene drives would significantly extend the spatial, temporal, and ecological bounds of industrial pest management to the scale of agro-ecosystems, providing a template of agro-industrial futures as plant biotechnology enters "an era of synthetic biology" (Power, 2021: 1). More specifically, this radical assemblage would bring fundamental and unknown changes in biodiversity, evolutionary dynamics and our relations with nature (ETC Group, 2016, 2018a; Medina, 2018; Power, 2021).

In sum, modern agri-biotechnogies are a paradigmatic expression of digital-molecular convergence, whose dynamics over the past 30 years or so have intensified and extended the structural fault lines and inequities of rural economy and society.

6

Between Physical Space and Digital Space: COVID-19, Platform Capitalism, and Changing Patterns of Food Provisioning

Several common themes recur as we move from upstream sectors to the digitalisation of food services, distribution, and consumption practices.[1] These include the rising ubiquity of digital platforms, consolidation in digital markets, increasing use of AI and robotics, and the worsening of long-standing structural fault lines: food poverty, health inequalities and the erosion of public safety nets. The COVID-19 pandemic cuts across these tendencies at random, amplifying early emerging trends with transformative force in some sectors while imposing difficult processes of adjustment on others.

As we explore how digitally mediated, platform-based food distribution and consumption practices are re-configuring the downstream food system, the COVID-19 pandemic is a constant and active presence. At the macro level, the pandemic is a disruptive force, exacerbating the multi-faceted global climate crisis but, in a sectoral context, disruptive change is a source of socio-technical opportunity, both of which heighten the existing inequalities and disparities in the food system. Buffeted by the vagaries of the pandemic, downstream sectors face an uncertain future as new actors challenge established ways of doing business.

The intensification of our digital engagements with food provoked by COVID-19, such as the remarkable rise of ordering-in meals at the expense of dining-out and takeaways, builds on the socio-technological infrastructures of the Internet, the mainstreaming of the World Wide Web in the mid-1990s, the mass-market diffusion of personal computers, and the expansion of social media platforms in the early 2000s. Gradually but inexorably, the digital has become entwined with "everyday food practices and experiences" (Lewis, 2020: 2), stimulating the proliferation of diverse "digital food cultures" (Lupton and Feldman, 2020). More specifically, "Over the past decade, the world of food, from grocery shopping and home cookery to restaurant going to food politics has been quietly colonised by the digital. Meanwhile, the digital has been invaded by all things food related" (Lewis, 2020: 2). Developments central to today's digital food practices also emerged gradually in the 2000s as part of this wider societal digitalisation, such as online grocery shopping and third-party home food delivery services. Two pioneers and current market leaders in app-based home food delivery – Takeaway and GrubHub – were established at that time, well before the explosive growth and takeover frenzy of 2020–2021.

Against this background, the COVID-19 pandemic of 2020 created the 'perfect storm', accelerating and deepening the entanglements between the digital and food services and consumption in a variety of ways. A key dynamic is the growing transition from the physical space of brick-and-mortar facilities to digitally mediated space, though it is simplistic to think that this transition is only one-way between two 'purified', entirely separate categories. We explore industry responses to potentially long-run, transformative changes in food shopping and eating habits that have been accelerated by the pandemic, and the structural changes that are likely to follow. This is also a way of saying that COVID-19 and its variants are an active, more-than-human presence in these new co-productions, assemblages, and networks.

The pandemic has given much greater salience to platform-based modes of food preparation and delivery, creating a new logistics infrastructure, which employs growing numbers of casual, piece-rate workers in the so-called 'gig' economy. Bohm et al (2020) characterise this process as 'Uber-isation': "to reflect how transporting food, once ordered, to the consumer becomes the new point of control". The emergence of 'dark' restaurants, 'ghost' kitchens and 'dark' grocery stores is the material expression of this transfer of power to online platforms, which are redefining the downstream food system.

On a wider canvas, the emerging logistics infrastructure, in ways as yet ill-defined, combines last mile delivery technologies that extend beyond food to packaged consumer goods and brick-and-mortar stores, whether 'dark', 'ghost', or traditional in form. The introduction of DoorDash Essentials, which combines deliveries from its own 'dark stores' and from partner convenience stores, such as 7-Eleven, across the US, and similarly replicated in the UK by Deliveroo Editions, illustrates how third-party meal delivery companies can exploit these wider retail opportunities. These have also attracted so-called 'quick commerce' firms making 'on-demand' home grocery deliveries from neighbourhood 'dark stores' stocking a limited inventory (Butler, 2021b). In short, if supply issues and last mile logistics can be resolved, *delivery app platforms are multi-purpose*, as readily applicable to packaged consumer goods as to food, as Amazon is now demonstrating in reverse since acquiring Whole Foods in 2017.

Retailers, online grocery shopping, and home food delivery

Over the course of the 1980s and 1990s, food retailers in the global North used M&As to raise levels of concentration and exert their market power to displace food manufacturing as the dominant force in the downstream food system. Grocery chains in EU countries were already highly concentrated by the 1980s and recent M&As have further consolidated the industry.[2] In the US, concentration was significant in regional and metropolitan markets but not at the national scale until the largest US and global retailer, Walmart, entered the grocery business in 1988. This coincided with the gradual easing of anti-trust regulation during the 1990s, which spurred a wave of M&As as other grocery chains, including Kroger and Safeway, sought to establish a national market presence (Howard, 2016; IPES-Food, 2017). In the years between 1997 and 2019, the share of the top four retailers in the US grocery market rose from 20 per cent to 43 per cent (Kelloway and Miller, 2019).

A second seismic change occurred in 2017, when the leading US e-commerce platform, Amazon bought Whole Foods Market for US$13.7 billion, prompting rival food retailers to accelerate their investment in online services, form partnerships with online grocery pick-up firms like Instacart, and take over online food delivery firms, as in the case of Target's acquisition of Shipt in 2017–2018. Amazon has since established a new grocery chain and food delivery service, Amazon Fresh, in order to extend its local presence nationally, supplementing the relatively small number of Whole Foods stores – fewer than 500 in the US – whereas Walmart has over 5,000 and Kroger has 2,700.

The larger national footprint of these chains has been a decided advantage in meeting the demand for online click-and-collect, kerb-side pickup services following the outbreak of COVID-19, and the introduction of emergency lockdown measures. Online ordering has become routine but customers have remained loyal to brick-and-mortar supermarkets and these pickup services developed as the preferred means of online grocery shopping during the pandemic (Mercatus/Incisiv, 2020; Field Agent Marketing, 2021). This preference for kerb-side, store pickup over home delivery apparently became stronger in the course of the pandemic, offering some reassurance to big-box retailers facing competition from on-demand, quick delivery 'dark store' grocers (Albrecht, 2021e).

The importance of location, aka prime real estate, in online grocery shopping is illustrated by this account from a journalist, who left his local store to become a 'Walmarter' due to the ease of its kerb-side pickup arrangements:

> I phone my order, schedule my pickup time, and when my groceries are ready they send me a notification on my phone. Part of that

> notification asks me to check in with the Walmart app to let the store know I'm on my way. As I pull into their parking lot, the app automatically recognizes that I've arrived, thanks to the GPS on my phone and Walmart's geo-referencing technology. (Albrecht, 2020a)

He adds, "Contactless is going to be the word of 2020, especially as it relates to food delivery and kerb side pickup, and geo-referencing is going to play an increasingly important part in that."

Although Amazon is now a competitive presence in online grocery markets (Koch, 2019), the rapid response of rival grocery chains and consumer loyalty have thwarted replication of its dominance of US e-commerce generally. As Kelloway (2020) observes, "despite much hand wringing, Amazon has yet to disrupt the grocery industry. Amazon Fresh delivery and Whole Foods controlled just 4 per cent of the grocery market in 2019". However, a consultancy report by Mercatus/Incisiv predicts that post-pandemic online grocery sales will rise from 3.5 per cent of total US grocery sales in 2019 to 21.5 per cent by 2025, suggesting that this rivalry will only intensify (Redman, 2020), as shown by the "increasing use of 'dark stores' that are not open to the public, and used only to fulfil delivery orders" (Howard, 2021: 43).

A rapidly growing version of the 'dark store' model is being propagated by a recent stream of start-ups that offer online grocery delivery at low cost within 10 or 15 minutes in selected, densely populated urban residential neighbourhoods. Start-ups operating 'dark stores' with a basic inventory of 2,000 or so grocery goods and a delivery radius of 1–2 miles include such firms as Weezy, Jiffy, Gorillas, and Getir in the EU, and Gopuff, JOKR, Food Rocket, and Fridge No More in the US (Albrecht, 2021c). These firms have enjoyed good funding rounds but were eclipsed in March, 2021, when Gopuff raised US$1.5 billion and a Berlin-based firm, Gorillas, which also operates in the US, raised US$290 million.[3] Gopuff reportedly makes app-based deliveries in 650 cities, has roughly 250 fulfilment or distribution centres located throughout the US, and bought the online wine and spirits delivery firm, Bevmo, for US$350 million in November, 2020. The impact of these ultra-fast delivery firms on the grocery market will depend on the adaptive responses of established retailers, the ability of these newcomers to scale, and whether or not modes of online home delivery continue their rapid growth as the pandemic evolves. We return to these developments in a later section.

For large, established grocery companies, accustomed to a slow growing, concentrated market with low margins, the projected expansion in online shopping offers a rare opportunity to gain market share and is likely to provoke further industry consolidation. Also, as widely observed, Amazon's decision to end its anomalous absence from this massive retail market can be seen as a long game to accumulate and monetise consumer data. These can be added to user data from Amazon Prime members, the growing popularity of the Amazon platform as a search engine, and other sources in order to improve the competitiveness of its brick-and-mortar stores in groceries and other retail sectors (Yglesias, 2017; Clinton, 2018; Fassler, 2018b). For Yglesias (2017) and others, a major factor in Amazon's competitiveness is that it continues to behave like a start-up whose primary focus is on re-investment and growth and net profit and dividend payments like its more traditional rivals.

With the rising popularity of online shopping, established grocery chains are adapting their physical space to the demands of digital services. This trend accelerated during the pandemic, and Walmart and other mainstream grocers are converting floor space in dozens of their stores into semi-automated warehouses, so-called 'micro-fulfilment centres', to speed up the processing and dispatch of customers' online orders (Bloomberg, 2019; CB Insights, 2020; Albrecht, 2021a). These distribution centres reveal a new openness of retailers to automation and robotics, reflecting market awareness of Amazon's e-commerce capabilities and rapid shifts in shopping practices caused by the continuing pandemic.[4]

In addition to the online delivery services offered by mainstream grocers, such as Walmart and Safeway, for example, and recent 'dark store' operators like Gopuff, many other actors also compete in this digital space, including firms with large, highly automated centralised warehouses and fleets of delivery vehicles, such as FreshDirect in the US, and Ocado in the UK. Another app-based grocery delivery model was introduced by the US tech start-up, Instacart, which has partnered with over 600 retail chains (Kelloway, 2021a), including Walmart, and provides 'personal shoppers' to complete online orders.[5] These 'shoppers' use their own transport to make deliveries, which has earned the company the sobriquet of the 'Uber of online grocery shopping'. Also like Uber, Instacart treats its 'shoppers' as independent 'contractors' or gig workers, without social benefits and other employment rights. This has led to stoppages, strikes and other labour disputes, as well as recent efforts by the United Food and Commercial Workers (UFCW) union to organise Instacart workers (Benner et al, 2020).

Meal kits are a further segment of the increasingly congested market for home food delivery, competing with online grocery shopping and app-based meal deliveries from brick-and-mortar restaurants and 'virtual' kitchens. This competition is acknowledged literally by the Canadian meal kit firm, Chef's Plate, whose website invites customers to "Say good-bye to meal planning and grocery shopping!" by joining its farm-to-table home delivery platform. Meal kits contain pre-proportioned fresh ingredients and recipe cards, and typically are sold on a subscription basis. The leading firms in the US – Blue

Apron, Hello Fresh, and Plated – were launched as start-ups in 2011–2012, and the two former companies are now publicly traded, while the grocery chain, Albertsons, acquired Plated in 2017, with plans to distribute meal kits through its 2,300 stores. A second major US grocery chain, Kroger, followed this lead in 2018 with its takeover of Home Chef. The world's largest food company, Nestle, was also attracted to this direct-to-the-consumer space in 2020, when it bought Freshly, which claims to deliver 1 million prepared meal kits nationally each week.

'Dark' restaurants, 'ghost' kitchens, and third-party, app-based food delivery platforms

Although Amazon's takeover of Whole Foods in 2017 and the 2020 pandemic compelled mainline grocery chains to adapt to changing consumer practices, in-store shopping, whether in person, online for home delivery or kerb-side pickup, retains its place as the bread-and-butter activity of food retailing. Indeed, supermarkets have fared relatively well in the pandemic despite investing heavily to improve their online shopping platforms, e-commerce capacity, and home delivery infrastructures.

In bleak contrast, the pandemic had a devastating impact on the restaurant business, forcing thousands of closures but also in many other, unforeseen ways, hugely inflating the gig economy with the growth of third-party ordering platforms and home meal delivery. This expansion of app-based platforms is re-drawing industry boundaries and provoking M&A activity on an unprecedented scale. This struggle for market dominance provides a vivid illustration of platform capitalism and its 'winner-take-all' power dynamics.

The leading meal delivery platforms, such as TakeawayJustEat, Grubhub, DoorDash, Deliveroo, and Uber Eats, exploit the digital interface between restaurants and their online customers, whose numbers have grown dramatically during the pandemic. For example, TakeawayJustEat increased its sales by 54 per cent in 2020 and other third-party delivery apps experienced similar rates of growth in sales revenues. Conversely, restaurants have been hard hit by the pandemic as mandatory lockdowns forced dining rooms to close, leading OpenTable, a US-based, international online reservation service, to predict that one in four restaurants in the US would go out of business (Marston, 2020a).

Restaurants have turned to take-out meals and online ordering, increasing their dependence on meal delivery apps and 'ghost' kitchen companies. These companies, such as Kitchen United and Cloud Kitchens, provide fully equipped kitchen facilities, variously known as 'ghost', 'dark', 'virtual', and 'cloud' kitchens, that are rented out to different restaurants to prepare meals for online delivery, available through app-based companies, adding a further dimension to the competition to become the gatekeeper between restaurants and their customers (Stoller, 2020). This category also includes the many restaurants that have converted their existing premises into 'ghost' kitchens and typically rely on third-party app companies to take their orders and make deliveries.

During the first months of the pandemic, particularly in the US, this reliance on meal delivery companies became controversial, as their dubious practices,[6] exorbitant commissions, and fees came to light, with independent restaurants and small chains being charged from 15 to 30 per cent or more (Jeffries, 2020; Kelloway, 2020b). In April, 2020, major cities across the US, fearing that restaurant closures would damage the local economy and reduce culinary diversity, imposed fee caps on third-party meal delivery firms.

However, even with this measure, as well as temporary aid from some meal app companies, restaurants and fast food chains still struggled to survive and began to evolve different responses to the crisis. These changes included encouraging customers to order directly by creating their own branded websites, developing in-house delivery services, and increasing their 'ghost' kitchen capacity to fulfil online orders. Fast food chains are now prioritising the re-modelling of their premises, reducing the size of dining room areas to release space for kerb-side delivery and pickup lockers using GPS and geo-referencing technologies, and to increase the number of drive-through lanes.

These pandemic-induced developments reveal that restaurants, third-party meal delivery companies and 'ghost' kitchen firms are embroiled in a zero-sum struggle to retain access to customers, their contact details, and food choices. In a Faustian bargain, restaurants are tempted to surrender their direct relations with customers in the hope of increased sales via apps and the availability of home delivery services. The fact that the commissions and fees paid by restaurants are the main revenue source of third-party meal delivery platforms emphasises the power asymmetry at the heart of their relationship.

On-demand, rapid delivery food services

As we have intimated, dine-in high street restaurants and fast food outlets are also threatened by *vertical integration* between food delivery companies and 'ghost' kitchen operators. This now increasingly well-trodden path was pioneered by the leading app-based home meal delivery platforms, including TakeawayJustEat, DoorDash, Uber Eats and Deliveroo, which are opening and operating facilities in competition with other ghost kitchen providers, such as Cloud Kitchen and London-based Karma Kitchen.[7] Deliveroo, for example, launched Deliveroo Editions, which are marketed as a localised hub or platform that hosts kitchens operated by restaurants producing meals for delivery only. "The result? A community-tailored collection of restaurants, providing the gold standard in food delivery" (Deliveroo Editions website, 2017).

The reality could hardly be more commonplace. One 'dark kitchen' operated by Deliveroo and branded among its Editions is located in Battersea, London, on an industrial estate, where "between intersecting railway tracks, a waste disposal facility, and a scaffolding company sits a nondescript industrial unit. Inside, a wide variety of international cuisine is being prepared, including Japanese, Lebanese, Greek and Italian" (Otway, 2020). This is one of "dozens of so-called 'dark kitchens' dotted across London" operated by Deliveroo, Uber Eats, and other meal delivery firms and variously located "in prefab structures, shipping containers and industrial units" (Otway, 2020). These company facilities are an extension of the gig economy employing low-wage, zero-hour workers with few benefits, cooking food in confined, socially isolated workspaces. We return to this question in more detail later.

As 'ghost' or 'virtual kitchens' proliferate, restaurant tech companies are developing single apps, such as Kitchen United's Mix platform and Crave Collective's virtual food hall, that allow customers to 'mix and match' when ordering meals from different virtual kitchens housed in the same building (Marston, 2021a). Such complex 'bundling' technology is a boon to virtual restaurants hoping to establish delivery-only brands (Bradshaw, 2019). The introduction of a single direct interface for online customers also adds to the competitive pressure on third-party food delivery companies and helps to explain why DoorDash, Deliveroo and others have entered the 'ghost' kitchen business.

The vertical integration in app-based food delivery noted previously is just the precursor of much greater cross-sector expansion as companies exploit the commercial opportunities created by the juncture between proprietary infrastructure, 'last-mile' logistics technologies, and the accelerating demand for contactless, convenient home delivery provoked by the COVID-19 pandemic. This potential is emphasised by Barwise and Watkins (2018: 28), who note that "Having a large digital platform requires massive infrastructure – servers, data storage, machine learning, payment systems, and so forth. Most of these have marked economies of scale and scope, enabling the business to take on other markets and rent out capacity to other firms, further increasing its efficiency and profitability."

These developments are particularly pronounced in the rapidly expanding on-demand, 'ten minute' delivery sectors, where the lines between the grocery, the ghost kitchen and the 'dark' convenience store are quickly becoming blurred as firms diversify away from their initial positions to serve all three categories, as exemplified by DoorDash, Gopuff, Food Rocket, and others (Marston, 2021b). Thus DoorDash launched its own national chain of delivery-only convenience stores, DoorMart, in 2020 and reportedly is investing in Gorillas, the speedy grocery delivery start-up, possibly with an eye to acquire a controlling interest. In turn, Gopuff is challenging

DoorDash's core activity by moving into online meal delivery by operating a chain of ghost kitchens (Albrecht, 2021f).

Vitaly Alexandrov, CEO of the San Francisco e-grocery delivery service, Food Rocket, offers insight into the thinking behind these cross-sector moves. He anticipates continued growth and further diversification, arguing that consumers will find on-demand convenience habit-forming. Interestingly, Alexandrov states that:

> We compete with Amazon. This may seem contradictory at first, given that Amazon wants to sell you everything (and) Food Rocket stores intentionally carry very few items. But again, Food Rocket is focused on your *daily* needs, not your every need. And it's fighting Amazon on the very battleground Amazon invented – speedy delivery. (cited by Albrecht, 2021d, original emphasis)

This versatility was recognised by Colin (2021) when he challenged the scepticism aroused by the listing valuation of Deliveroo on the eve of its flotation on the London Stock Exchange in April, 2020:

> It's already apparent that Deliveroo has a broader scope than meal delivery. The company's main asset, after all, is the infrastructure it's been deploying over the years, connecting couriers with supporting technology and collecting data at every turn
> [...]
> now the same infrastructure can be used for much more than that: ordering meals prepared in 'cloud' kitchens, ordering groceries online, perhaps even delivering prescription drugs
> [...]
> The stakes are high: nothing less than upgrading last mile logistics and rewiring dense European cities so as to increase the level of convenience for all.

Such crossover possibilities beyond food services are exemplified by the Singapore-based firm, Grab Holdings Inc., which has diversified from its origins in 2012 as a ride-hailing app into food and grocery delivery, insurance, payments, e-wallets, and finance (Ruehl and Palma, 2021). Grab's 'sell-you-everything' model via its 'super-app' is familiar in Asia, but "The easiest way to describe Grab to a US audience is to say that it's like Uber, Uber Eats and Apple Pay all in one" (Lachapelle, 2021). Grab is hoping to build on the diversified e-commerce models pioneered by Amazon and Alibaba, complementing the various forms of last mile home delivery, an individually slight yet cumulatively significant sign of the digital transformation of the downstream food system and retail commerce in general.[8]

Platform capitalism, consolidation, and the gig economy

Before we consider the recent wave of consolidation in the third-party food delivery sector of the platform economy and the casualisation of its work force, Barwise and Watkins (2018) offer a more general perspective on two-sided or multisided online markets. Using the restaurant industry as illustration, the network effects in these markets are 'indirect' because "the value to participants in each market (e.g., diners) depends on the number of participants in the *other* market (e.g., restaurants), and vice-versa. Once a platform dominates the relevant markets, these network effects become self-sustaining as users on each side help generate users on the other" (Barwise and Watkins, 2018: 26–27, original emphasis). In a generalisation that describes today's app-based food delivery companies perfectly, these authors state that:

> *All businesses that depend on indirect market effects face the 'chicken-and-egg' challenge of achieving critical mass in both or all the key markets simultaneously.* Until the business reaches this point, it will need to convince investors that early losses will be justified by its eventual dominance of a large and profitable multisided market. (Barwise and Watkins, 2018: 27; my emphasis)

This vital trajectory towards *critical mass* helps to resolve the apparent contradiction that the leading app companies in both the US and the EU are continuing to make losses, despite the surge in demand triggered by the pandemic. Booming sales were not enough to prevent Deliveroo, for example, from making losses of £224 million in 2020, and the same is true of its European rival, TakeawayJustEat, and leading firms in the US, such as Uber Eats and DoorDash (Jeffries, 2020). *A major underlying issue is the failure to cover home delivery costs, even when using gig workers for the 'last mile'*: "In 2019, analysts from the investment firm Cowen estimated that Uber Eats was losing US$3.36 on every order and would continue to lose money on every order for the next five years" (Jeffries, 2020). This persistent flaw in the business model is now being magnified by growing pressures to extend employment rights to gig workers, as discussed later.

While recognising the challenges of multisided markets, the elusiveness of profitability amid soaring sales revenues raises issues about the viability of the operating model of food delivery app platforms. These are most closely interrogated when firms file for an IPO, as shown in the cases of DoorDash and Deliveroo. In a business with wafer-thin margins, the model followed by these firms is based on exploitation of two of its integral elements: the restaurant industry – an equally low-margin business, also with questionable employment practices – and its gig economy 'contractors'. These underlying realities tend to be obscured by the razzmatazz of seemingly constant M&As in a textbook demonstration of the relation between platform dynamics and consolidation.

The key factor here, as Barwise and Watkins (2018) observe, is plentiful venture capital funding to finance predatory, loss-making 'unicorns', and takeovers are an expedient way to increase the customer base and so attract more restaurants to the platform in the hope that greater size will deter new entrants, raise margins, and lead to profitability. This is the growth path taken by the leading US firms in 2020 – DoorDash, Uber Eats, Grubhub, and Postgates – which rose to prominence through successive rounds of takeovers of local rivals to gain dominance in their respective metropolitan markets of San Francisco, Miami, New York, and Los Angeles (Kelloway, 2020b; Stoller, 2020). TakeawayJustEat and Deliveroo followed a similar route to the top in continental EU markets and the UK.

The surging demand for home meal delivery following the outbreak of COVID-19 in 2020 brought the struggle for platform dominance among these loss-making companies to a head, and it is now being fought out on a global scale. In apparent awareness of the 'winner-take-all' dynamics in these digital markets, a spate of takeovers and IPOs occurred in 2020–2021. This wave of consolidation was initiated by the Dutch restaurant platform, Takeaway, which acquired UK-based meal delivery company, Just Eat, with an all-share offer of £5.9 billion in January, 2020. Six months later, in a successful move to forestall the ambitions of Uber Eats, TakeawayJustEat completed the acquisition of the US firm, Grubhub, again with an all-share offer of US$7.3 billion, making the new company the largest in the world outside China.[9] Having failed to acquire Grubhub, Uber Eats responded by buying Postgates for US$2.65 billion in November, 2020, also in an all-stock transaction, leaving three firms controlling 98 per cent of the US market (Kelloway, 2020b).

Although analysts questioned its high valuation and post-pandemic prospects, DoorDash completed its IPO on the New York Stock Exchange in December, 2020, achieving a surprisingly high market capitalisation of US$72 billion, despite losses of US$667 million in 2019 and US$149 million in the first three quarters of 2020, richly rewarding the owners, bankers, and venture capitalist investors in this 7-year-old start-up. Commenting on the overpriced IPO of this lossmaking company, one market analyst suggested that "This is Silicon Valley selling public markets an asset at a huge premium, and they're going to laugh all the way to the bank"(David Trainer, Market Constructs, cited by Roberts, 2020).

The British firm, Deliveroo, in which Amazon acquired a 16 per cent holding, aroused similar misgivings about its valuation following its failure to turn booming sales revenues into consistent profitability, recording losses of £224 million in 2020. On the eve of its IPO on the London Stock

Exchange on 31 March, 2021, Deliveroo acknowledged these concerns by cutting its flotation share price from 460p to 390p, reducing its expected capitalisation from £8.8 billion to £7.6 billion. Nevertheless, a complex of factors relating to the nature of the app-based industry combined to make this a disastrous flotation – 'Floperroo' – as Deliveroo's share price fell by 26 per cent in first day trading, reducing its expected valuation by roughly £2.5 billion.[10]

Unlike DoorDash, Deliveroo's reliance on gig workers was a major concern in the run up to its flotation as this question had already attracted increasing scrutiny in the UK and EU markets, notably in Italy and Spain, where the firm was facing on-going legal challenges.[11] In turn, this scrutiny increased fears of imminent public regulation to re-classify independent contractors as employees, a move that would seriously jeopardise, if not break, the business model of Deliveroo and its app-based rivals. A recent report by the Bureau of Investigative Journalism in the UK found that the majority of Deliveroo's self-employed couriers earned less than the UK adult minimum wage, prompting one analyst to characterise this issue as a 'ticking bomb' (Strauss et al, 2021).

This context is framed by Uber's recent decision to accept a ruling by the UK Supreme Court and grant certain social benefits to its taxi drivers and recognise them as employees, which has put intense pressure on Deliveroo and other UK app-based food delivery platforms to follow suit. Ethical concerns about the treatment of its 'contractors', as well as the regulatory risks of an enforced change to an employee-based business model, persuaded several leading UK investment funds, including Legal & General and Aviva Investors, not to take part in the Deliveroo IPO. These events, and the recent dismissal of an appeal by the gig economy taxi firm, Addison Lee, against a landmark employment tribunal decision recognising its drivers as workers eligible for the minimum wage, appeared to indicate the 'direction of travel' in the UK (Butler, 2021a).[12]

The regulatory challenge to the business model of gig economy platform companies in the EU became concrete in late 2021, when the European Commission published draft legislation to reclassify gig workers as employees eligible for employment benefits. These proposals, if adopted, would be applicable only in the 27 EU member states but some observers suggest that a 'Brussels demonstration effect' may lead to their extension to the UK and other non-member countries (Rankin, 2021).

California's Proposition 22

In the US, the direction of travel is more difficult to foresee, more tortuous, perhaps reflecting the more openly 'entrepreneurial' political economic culture. App-based platforms are engaged in a continuing battle to resist legislative efforts to re-classify their gig workers as employees. In a recent round, platform capitalists appeared to have gained the upper hand following the victory in November, 2020 of Proposition 22. This ballot initiative was a reaction to the passage of California Assembly Bill 5 (AB5), later upheld by a Superior Court judge in August, 2020, requiring gig economy companies to comply with AB 5 and re-classify their gig workers as employees eligible for social benefits, including the minimum wage, paid holiday leave, and health insurance. A coalition of gig worker platforms, including Uber, Lyft, DoorDash, and Instacart, "spent roughly US$200 million on the ballot measure, making it the most expensive in California history" (Marston, 2020b). Having apparently seen off the threat of AB5 to its business model, these companies turned their attention to defeating worker re-classification initiatives in Massachusetts, Illinois, and other states.

However, the ballot victory of Proposition 22 may well prove to be hollow as an Alameda County Superior Court Judge declared in August, 2021 that it is "unconstitutional and unenforceable". In ruling on a lawsuit filed by the Service Employees International Union, Judge Frank Roesch ruled that the legislation illegally "limits the power of a future legislature to define app-based drivers as workers subject to workers' compensation law" (cited by Lyons, 2021). At this writing, the corporate coalition behind Proposition 22 is appealing this judgement and the legislation meanwhile remains in force, pending the outcome of the appeal process.

Judge Roesch also went to the heart of objections against Proposition 22 and its attempt to write gig employment practices into state law, finding that "It appears only to protect the economic interests of the network companies in having a divided, un-unionised workforce which is not a stated goal of the legislation" (cited by Lyons, 2021).[13] This finding is given an academic gloss by Veen et al (2020), who note that the platform economy is variously "viewed as a continuing trajectory of neoliberalism (Zwik, 2018) … with some characterising platform work as digital Taylorism (Cherry, 2016) and others as post-capitalist (Pelicca-Harris et al, 2020)". Whatever the conceptual label, however, the constants of platform work organisation are undisputed: minimal, if any, employment benefits, the transfer of employment risks to the workers via piece rates, and uncertainty about the availability of work, resulting in the extreme precariousness of this form of low-income employment. The conditions encountered under digital labour platforms are described in personal terms by Sarah Mason, who worked for the ride-hailing company, Lyft, and experienced the effects of the opaque algorithmic management of its workers directly (Mason, 2018).

As the vagaries of Proposition 22 and the Deliveroo IPO reveal, the regulatory trajectory of gig economy platforms is still in flux, although some recent initiatives are now clarifying the situation. As just noted, the success of the coalition behind Proposition 22 in portraying their couriers as entrepreneurs, legitimising their status as independent 'contractors' (Veen et al, 2020), has been called into question by judicial decisions in California.[14] This entrepreneurial framing has less traction in the EU, as shown by the European Commission's proposals to reclassify gig workers, building on previous piecemeal legislation by some member states, which will establish a unified regulatory structure for employment practices in the rapidly expanding platform economy in the EU (ILO, 2021; Rankin, 2021). With the EU moving to curb exploitative gig labour practices and California's Proposition 22 mired in the judicial appeals process, will venture capitalists still have the appetite to fund rapidly growing but loss-making app-based companies in these immature sectors, and engage in the type of profit-seeking behaviour that Stoller (2020) associates with 'counterfeit capitalism'? The answers to this question should emerge in the course of 2023.

COVID-19 as socio-natural collective

So far in this chapter, we have treated COVID-19 as a socio-technical catalyst, accelerating existing trends and uncovering latent possibilities in the digitalisation of food provisioning and consumer behaviour, and notably the rise of app-based delivery platforms.[15] More significantly, however, COVID-19 is a mutable socio-natural collective of non-human and human entities: emerging in a Wuhan wet market to become a global health crisis, transmuted into an economic crisis, a food crisis, and now taking on all these forms simultaneously. The pandemic has caused a severe economic crisis globally, revealing the frailty of public safety nets, exposing long-standing social inequalities, and magnifying racial and ethnic injustice. While it is difficult to generalise at this level,[16] COVID-19 engendered economic crises in the UK and US in 2020 that quickly became crises of food insecurity, encompassing the 'old' food poor, but also rapidly rising numbers of the 'new' food poor (Caplan, 2020a).

The severity of the COVID-19 food poverty crisis in the UK is exacerbated by the dismantling of social welfare safety nets since the 2008 financial crisis, leading to the privatisation of public welfare, or what Caraher (2020) calls its 'charitisation'. After more than a decade of economic austerity policies, the UK Food Standards Agency estimated that in 2019, *before the pandemic*, roughly 20 per cent of adults were food insecure (Caplan, 2020b). According to Emma Reeve, chief executive of the Trussell Trust, Britain's largest food aid charity, with some 1,200 affiliated food banks, "What we are seeing this year is people's financial resilience eroded, people going into debt. More people teetering on the edge, and this system is not equipped to catch them all" (cited in Butler, 2021). These points are supported by Caraher and Furey (2021), who note that "it is clear that the COVID-19 crisis has resulted in higher levels of indebtedness, with impacts on food poverty", particularly among low-income households. The pandemic has reduced life expectancy at birth in England, with "the most pronounced effects ... in vulnerable groups, where inequality has widened" (Longevity Science Panel, 2021).

In the US, unemployment claims reached levels reminiscent of the 1930s and, together with nation-wide school closures, placed unprecedented demands on food banks and other charitable sources. The Institute of Policy Research at Northwestern University found that food insecurity had reached record levels, with one in four Americans lacking enough to eat in mid-December, 2020 (Lakhani, 2021). In the words of Dr Leana Wen, professor of public health at George Washington University, "COVID has amplified existing disparities in education, food insecurity, unstable housing and health outcomes for a whole generation of children. ... Children have borne the brunt of the lack of action to contain COVID-19 and the failure to prioritise schools" (cited in Lakhani, 2021). Research in both the US and UK also revealed the racialised contours of economic inequality and food poverty (Caplan, 2020b; Power et al, 2020; Lakhani, 2021).

COVID-19 is a pandemic of many facets. It has created a range of economic opportunities for both established food retailers and a host of start-ups developing contactless food delivery platforms. Other food providers, including restaurants and fast food outlets, have struggled to cope with COVID-induced socio-technical change. Beyond these 'parochial' sector concerns, COVID-19 lockdowns, anxiety, and uncertainty have caused social breakdown, manifest in the severe economic hardship and food insecurity suffered by millions globally, simultaneously exposing the deep inequalities and injustices endemic to both economic and food systems alike.

Analysis and conclusion

This chapter has conveyed the turbulent and protean nature of the app-based home delivery sector as firms respond to the rapidly expanding demand for contactless convenience provoked by the COVID-19 pandemic. This shift to online shopping is also accelerating in the more mature grocery industry, overcoming initial reluctance to digitalise operations, and given added momentum by Amazon's recent entry. Overall, this chapter leaves no doubt about the relentless digitalisation of food services, shopping and consumption, although this process depends on the physical, 'brick-and-mortar' spaces of ghost kitchens, fulfilment centres, supermarkets, 'dark' stores and restaurants, and the infrastructure requirements of digital platforms. These spaces are also vitally important to firms' efforts to use their digital presence and home delivery logistical systems to diversify into other retail channels, as exemplified in their different ways by Amazon, DoorDash, Gopuff, and Grab.

We also saw that the app-based home delivery industry is grounded in the exploitative work practices of the gig economy that frame Deliveroo's biker couriers and Instacart's thousands of 'shoppers' as independent, self-employed contractors, even small-scale entrepreneurs (Veen et al, 2020). This substandard treatment of their workforce is fundamental to the 'last-mile' logistical systems on which the reproduction of companies in this segment of the platform economy depends. However, this exploitative form of capitalism is now being challenged, as shown by the changing fortunes of Proposition 22 in California and the prospect of a coherent, systematic approach to its regulation in the EU.

The momentum of digitalisation and the crucial importance of 'last-mile' logistics to profitability can be illustrated by one closing example: the UK-based online retailer, Ocado. This company announced recently that it is partnering with the Oxford University spin-out, Oxbotica, to develop autonomous, self-driving vehicles to make home deliveries, "bringing a world where online grocery orders are picked, packed and delivered entirely by robots one step closer" (Sweney and Wood, 2021). This move to a robotic workforce is expected to reduce labour costs: "the cost of moving orders within Ocado's warehouses amounts to 1.5% of UK sales, whereas ... the cost of 'final-mile delivery' is about 10% of sales. Labour represents about half of these costs" (Sweney and Wood, 2021). Interest in self-driving delivery solutions is also rising in the US, where Domino's and Chipotle are partnered with the autonomous vehicle start-up, Nuno, and Walmart with two other start-ups: Cruise for autonomous last-mile delivery, and Gatik over medium-distances (Albrecht, 2021b).

The development of autonomous vehicles for online food delivery resonates with the fully automated industrial farms operated by 'intelligent' machines as envisioned by Dodge (2019), for example. Such observations at each end of food supply chains also point to the advantages of a systemic approach to food and its many environments, techno-scientific, ecological and cultural. It may be objected that the premise of digital-molecular convergence has less purchase on downstream sectors of the value chain, but this would be to ignore the ubiquity of convergent technologies in fractionating the ingredients for the fabricated, highly processed foods, not least meat analogues, which dominate contemporary consumption.

There is broad consensus that the COVID-19 pandemic accelerated the diffusion of digital technologies and practices in 2020–2021. If the inequitable global distribution of vaccines continues – 'The virus anywhere, the virus everywhere' – new variants are set to emerge, particularly in less developed countries, for years to come. In other words, in the absence of global leadership on the question of fair access, the pandemic will shape political economies in both rich and poor countries, forcing actors in the downstream food system to respond and adapt to successive waves of infection. As a corollary, trends heightened by the pandemic in 2020–2021 are likely to become permanent structural features. New variants of COVID-19 and their mutations, some of global significance, are taking us through the Greek alphabet and raising the spectre of further disruptive lockdowns. Thus the B.1.1.529 or Omicron variant, encountered in southern Africa with 32 mutations in the spike protein, undermined hopes of a full 'return to normality' in rich countries in 2021, despite the success of vaccine booster drives. In short, the probability of new, significant variants creates a continuing state of uncertainty and the threat of recurrent crises, marked by severe economic recession, poverty, food insecurity, and widening socio-economic inequalities.

On this prognosis, in-house restaurant dining is likely to lose further ground to online meal delivery, meal kits, and home cooking. Structural change in the grocery industry is more difficult to assess, since, despite the inroads made by app-based, 'dark store' delivery, the powerful big box retailers are adapting to this challenge by developing their own online and on-demand services. Furthermore, competition may give way to collaboration and new trajectories of consolidation. Thus, for a trial period, Tesco UK will partner with Gorillas to make on-demand, '10-minute' deliveries from 'dark' micro-fulfilment centres set up within its supermarkets (Butler, 2021d).[17] The wider adoption of this model by other established incumbent retailers would represent an intriguing hybrid response to the COVID-19 pandemic and the trends it has generated.[18]

7

Conclusion: Continuities in Change and Lost Opportunities

In a book on technological innovation and change, it is surprising to find so many strong continuities on multiple scales: from the global food system and agro-industrial sectors to competitive 'treadmill' innovation at the farm level. Looking back over individual chapters, the shifts in space, time, scale, and social practice that co-produce new socio-spatial relations – the high-resolution, miniaturised digital 'grids' of PA, or gene driving at agro-ecosystem scale – are being accommodated within the hegemonic industrial agro-food system. Indeed, the longevity of this political economic paradigm is the central and most striking continuity of all. To understand these historical continuities, we have drawn on evolutionary economics, and particularly the concepts of lock-in and path-dependence.

The structural changes generated by this model, including the loss of rural livelihoods, farm indebtedness, land consolidation, corporate concentration, and environmental degradation, for example, are equally long-standing and have captured the attention of past generations of scholar-activists, such as Rachel Carson, author of *Silent Spring* (1962), and social scientists, particularly with the emergence of critical rural sociology and the 'new' political economy of agriculture in the 1980s (Buttel and Newby, 1980; Newby, 1983; Friedland et al, 1991). These deep-seated trends and the underlying path dependencies have intensified in recent decades, as we saw in Chapters 2, 3, and 5, drawing on the research of John MacDonald and colleagues, the IATP (2020), Philip Howard (2016, 2021) and reports by the ETC Group.

The critique of the IATP (2020) has particular salience here with its indictment of US farm policy for its serial failure to arrest, let alone reverse, the social injustices exacerbated by these trends, extending from the international farm crisis of the 1980s. Perhaps most significantly, the IATP highlights an enduring continuity in the ethos of farm policy, articulated by Secretary of Agriculture, Earl Butz's mantra in the 1970s and reiterated by a successor, Sonny Perdue, in 2018, of 'Go big, or get out', expressing a profound indifference to the future of the family farm, rural employment, and rural communities.

The view that PA represents evolutionary rather than radical, disruptive change, which reinforces the hegemonic industrial paradigm, was advanced several decades ago by Wolf and Buttel (1996), and is strongly supported by recent analyses, as discussed in Chapter 3. Corporate digital service platforms marketing proprietary seeds and agro-chemical inputs are the organisational expression of this continuity which, despite 're-scripting', is commoditising farmers' experiential knowledge. The John Deere vision of 'Farmer Terry' in his living room watching robots 'at work' in his fields says it all (Bronson, 2018). Cochrane's (1979) concept of the treadmill of competitive innovation is the perfect metaphor of technological lock-in and path-dependence.

Agro-biotechnologies similarly follow a well-established trajectory from first generation innovations in the 1990s to later iterations of genomic sequencing, gene editing, synthetic biology and gene driving (Benbrook, 2018; Bronson, 2018, 2019; Clapp, 2021). These iterations reinforce the commercial axes of the chemical-intensive industrial model, but the potential use of gene drives for pest control as implied in the foundational patents foreshadows a radical transition. It would mark a quantum change in the scale of genetic intervention from individual species to the agro-ecosystem level, so-called 'ecosystem engineering', further undermining agricultural biodiversity and threatening the integrity of the biosphere. As we argued in Chapter 5, this scalar shift is occurring at a time when regulation of agri-biotechnologies has effectively been captured by a handful of life science corporations.

As we have seen throughout this book, the convergence of digital and genetic technologies has not significantly disturbed the corporate architecture of the agro-industrial model. Long-established 'heritage' companies have adapted well to the organisational shift to digital platforms and the new forms of capitalist competition that have accompanied Big Data technological innovation in PA and alternative proteins. In each case, M&As and related entry strategies have been decisive, and this 'predatory' activity has brought processes of corporate concentration and consolidation in food systems into renewed focus.[1] This builds on ground-breaking research by William Heffernan and colleagues at the University of Missouri (Heffernan et al, 1999) and is advanced in recent contributions by Philip Howard, notably on the consolidation of the 'seed-chemical complex' from several dozen companies in the 1970s to its current incarnation as the 'Big Four'.

The established corporate architecture looks less secure in downstream food provisioning in the wake of COVID-19, which has brought dramatic changes in consumer shopping behaviour and ways of eating, and accelerated the societal penetration of digital app-based platforms. Start-ups are taking online purchasing in novel directions, posing new challenges to traditional big-box retailers and in-restaurant dining. Companies that have devised last-mile infrastructure for app-based home food delivery using gig economy couriers are now diversifying their activities into groceries and convenience shopping, meeting on-demand orders in short delivery times from their 'dark stores' and ghost kitchens. Although supermarkets are developing their own on-demand quick delivery services, this immature sector displays a strong 'winner-take-all' dynamic, evidenced by loss-making start-ups, abundant venture capital funding, highly inflated valuations, and numerous IPOs. This flux is likely to continue as new provisioning practices stimulated by COVID-19 and reinforced by its variants, become embedded in foodways.[2]

Sustainability transition?

The socio-technical innovations examined in this book are contextualised by global climate change and the imperative need to reduce GHG emissions in the global agro-food system. A recent FAO study presented at the COP26 climate conference held in Glasgow in November, 2021, estimated that the global agro-food system contributes 31 per cent of total anthropogenic emissions (FAO, 2021). Nevertheless, despite urgent, repeated calls by the IPCC for a global sustainability transition to reduce human influence on the climate, most recently in its Sixth Assessment Report (IPCC, 2021), there is scant evidence that industrial agro-food systems in the Global North are committed to a transition pathway to a sustainable, low carbon future. Nor should the scaling up of the emergent alternative protein industry, shadowed by Big Food 'protein conglomerates', be confused with broadly-based, system-wide paradigm change.

That said, it is misguided to disparage a dietary transition away from conventional meat production to plant-based proteins as little more than a 'techno-fix'. While discursive claims can easily be exaggerated, it would bring welcome savings in carbon emissions, release natural resources, and help to mitigate global warming (Boston Consulting Group/Blue Horizon Corporation, cited by Carrington, 2021). A dietary shift towards alternative proteins would be one among hopefully many steps taken to avoid extreme global warming.[3]

In comments on the failure to pursue localised, bottom-up, equitable, and more resilient, sustainable food system trajectories, critical agro-food scholars draw an analytical parallel between industrial concentration and the specialisation of agricultural R&D on a narrow range of chemical-intensive, mono-cultural, and internationally-traded commodity crops, particularly maize (Mooney, 2015; ETC Group, 2018b; IPES-Food, 2017), Essentially the same point is made by Heinemann (2020: 113; my emphasis), who notes that agricultural research systems in developed economies increasingly have been *devolved* to the private sector "and therefore to the structures and incentives that drive the private sector. The ultimate market concentration that results, *reduces options for agriculture in both developed and developing countries* because modern biotechnology has mainly served green revolution-type breeding demands to fit high input and uniform agro-ecosystems." Research contributions discussed in Chapter 3 stressed how path dependent techno-scientific knowledge and design practices privilege and normalise hegemonic farming styles (see Bronson and Knezevic, 2016; Bronson, 2018, 2019).

Counter-futures, the United Nations Food Systems Summit and 'anticipatory politics'

The current and imminent dangers of climate change as critical 'tipping points' draw ever closer and planetary resource boundaries are breached have intensified calls to transition to an agro-ecology model to address the food systems crisis in equitable, socially just, and ecologically sustainable ways. The international consultative process held between 2004 and the release in 2009 of the International Assessment of Agricultural Knowledge, Science and Technology for Development (IAASTD) (2009) report, 'Agriculture at a crossroads', marks a major watershed in efforts to change the narrative and discursive framework to support a holistic, 'food systems' approach to agriculture and food. "The 2009 IAASTD report was the forerunner of many subsequent reports that bring together the silos of agriculture, health, climate, social equity and economics" (Swinburn, 2020: 133). Later reports 'in the spirit of IAASTD' (Herren, 2020) have been published by social movements, civil society organisations and inter-governmental agencies, including United Nations Environment Programme (UNEP), United Nations Conference on Trade and Development (UNCTAD), and the FAO. Following reforms in 2009 to give wider representation to civil society and private sector groups, the United Nations Committee on World Food Security (CFS), its High Level Panel of Experts on Food Security (HLPE), and IPES-Food have made landmark contributions to these discursive struggles to persuade policymakers, scientists, and agribusiness of the urgent need to reform the global food system and undertake a sustainability transition.[4,5]

Participants in the IAASTD process believe that "a new food system narrative has been firmly established over the past decade" (Haerlin, 2020: 17), with agro-ecology as its unifying framework, understood "both as a social and cultural concept and as a set of agriculture and food system practices" (Herren, 2020: 19). Many contributors to later reports inspired by IAASTD share this belief but, as one protagonist in this discursive field observes, discourse is not the whole story: "My over-arching sense is that in the past decade we have made considerable strides at the levels of paradigms, concepts, rhetoric and global commitments but the policy action on the ground has remained patchy and sluggish – *far too sluggish for the urgency that the global food system crisis demands*" (Swinburn, 2020: 130; my emphasis).⁶

This sense of policy inertia or drift was accentuated recently by growing concern that past narrative gains are now being actively eroded as corporate interests co-opt and re-invent the IAASTD discourse of food systems transformation and sustainability transition (see also Herren, 2020). Supporters of transition to an agro-ecological paradigm fear that this pathway is now being deliberately blocked by a coalition of agribusiness corporations, powerful OECD states, the elitist corporate think-tank, the World Economic Forum (WEF), private philanthropists, notably the Gates Foundation, the Consultative Group on International Agricultural Research (CGIAR), and other institutional advocates of industrial 'green revolution' technologies. These fears were grounded in efforts by agribusiness and allied interests to capture and privatise global food governance by controlling the agenda, participation, and organisational structures of the United Nations Food Systems Summit (UNFSS). This was convened in 2019 to set out the path to achieve the UN's Sustainable Development Goals (SDGs) and held in New York in September, 2021.

This disquiet intensified when the United Nations Secretary-General, Antonio Guterres, appointed Dr Agnes Kalibata, then President of the Alliance for a Green Revolution in Africa (AGRA),⁷ as Special Envoy responsible for the UNFSS. Critics of the planning process objected *inter alia* to the strategic partnership agreement with the WEF, the lack of transparency and accountability, the *de facto* exclusion of social movements and indigenous peoples, neglect of alternative forms of knowledge, conflicts of interest, and the failure to prioritise human rights, particularly the right to adequate, nutritious food (Canfield et al, 2021; ETC Group, 2021a; IPES-Food, 2021a, 2021b).

Adoption of the corporate concept of 'multi-stakeholderism' to legitimise participation in the UNFSS was seen as a stratagem to subvert the institutions of international public governance of global food systems, notably the FAO, the CFS and the HLPE, its science-policy interface (SPI), and other Rome-based inter-governmental agencies. This move to place private interests at the centre of food systems governance follows the WEF's global 'reset' agenda to establish post-pandemic 'multi-stakeholder' governance, described by critics as an 'artificial new multilateralism' (IPES-Food and ETC Group, 2021). It is an expression of the 'stakeholder capitalism' promoted by the WEF that effectively reproduces existing power structures (see also Canfield et al, 2021).

The UNFSS process also can be seen as a form of 'anticipatory politics', whereby powerful actors attempt to pre-empt techno-scientific futures in order to manage the present and maintain the status quo (Anderson, 2010; Jeffrey and Dyson, 2021). This perspective aptly describes how the UNFSS was manipulated *to ensure that certain technocratic forms of knowledge dominate the highest levels of global food system governance and define visions of agro-food futures.*

These growing concerns were heightened by the sidelining of the Rome-based multi-lateral agencies of global food governance in the formation of the Summit Scientific Group, its narrow disciplinary composition, and close links to advocates of an 'IPCC for Food'. If this proposal were to be adopted, the new inter-governmental panel would become the leading scientific authority on food system transformation, displacing the FAO/CFS/HLPE structure.[8] Several vocal proponents of an 'IPCC for Food' as the new science-policy interface were recruited as members of the Summit Scientific Group (IPES-Food, 2021a), which was dominated by hand-picked natural scientists and economists to the exclusion of other social science disciplines, agro-ecologists and experts on indigenous knowledge and human rights: "[The] Science Group is acting as a gatekeeper for determining the meaning and boundaries of 'science'" (Montenegro de Wit et al, 2021a).[9,10]

Techno-centric path dependence redux

The processes and events surrounding the organisation of the UNFSS bring the central arguments of this book full circle. Schematically, the UNFSS was commandeered to endorse the industrial techno-scientific model at the global scale by undermining public food system governance and introducing corporate, private interest 'multi-stakeholderism' in its stead. An 'IPCC for Food' as the new science-policy interface would ensure the institutional and epistemic hegemony of orthodox establishment perspectives on food system transformation. Agro-ecological approaches and scientific practices would continue to be marginalised, together with openness toward different forms of knowledge, food sovereignty and human rights.

A further chance to re-calibrate food system trajectories was spurned at COP26, the global climate conference held in Glasgow in November, 2021. Remarkably, despite generating fully one-third of GHG emissions, with industrial livestock and dairy production alone contributing 14–15 per cent, the food system was largely peripheral to the COP26 agenda and political negotiations (Askew, 2021a, b; Weston and Watts, 2021). Moreover, those national climate plans submitted to COP26, known as Nationally Determined Commitments or NDCs, that did include food system goals, conceived food policy in narrowly 'productionist' terms and so gave priority to on-farm innovation.

This focus translates into 'climate-smart' technologies, notably those deployed in PA incorporating gene editing and synthetic biology (ETC, 2021b). For Feehan (2021), who notes that the share of pre-and post-production activities in food system GHG emissions has been rising since

1990, the prioritising of production in NDCs represents a "colossal missed opportunity" by G20 governments to support a dietary transition from farmed meat and dairy products to plant-based substitutes. Surveying the submissions by G20 countries, she laments that "When it comes to dietary change in NDCs, there's an eerie silence."

Proceedings at the UNFFS and COP26 ostensibly were devoted to articulating a global transition to low carbon futures. In practice, however, food system proposals at each conference demonstrated yet again that technocentrism remains the touchstone of establishment scientific and policy approaches to sustainability transition. This book has identified many of the ills and harms caused by this techno-scientific, political economic, and epistemic path dependence, which underpins the continuity of the industrial agro-food system. Put simply, the dominance of highly concentrated, transnational conglomerates or Big Food over this political economy is antithetical to the urgent and comprehensive articulation of sustainable, socially equitable, and inclusive responses to the converging crises of climate change (see also Bene, 2022).

Ideally, the UNFFS would have built bridges between techno-centric and agro-ecological approaches,[11] overcoming entrenched resistances and tensions to find common purpose in the challenge to reduce the hugely damaging contribution of the food system to global warming. This implies willingness to abandon the 'silos' of techno-centrism and eco-centrism, and openness to hybrid strategies to achieve the SDGs and limit global heating to 1.5 °C above pre-industrial levels. Unfortunately, as we have seen, the proceedings of the UNFFS were marked not by collaboration and compromise but acrimony and exclusion. Now, time is of the essence if the planet is to avoid catastrophic heating and secure a liveable and sustainable future for all (IPCC, 2022).

Despite increasingly pressing calls for decisive action to mitigate this existential threat, the political will and multilateral institutional leadership needed to undertake the rapid, coordinated, root-and-branch transformation of the political economy of food systems are still conspicuously lacking. As Bene (2022) also concludes, the "Great Transformation of food may not happen". If not, this impasse will virtually rule out the possibility of reaching the IPCC target to reduce GHG emissions by half by 2030 in order to keep global warming below 1.5 °C.[12]

The key premise of this book is that the current wave of innovation driven by the convergence of digital and molecular technologies has been contained within the hegemonic industrial model of agriculture and food. Locked into long-established parameters, innovation has brought incremental, evolutionary change following trajectories articulated decades earlier, and heavily supported by public agriculture and food policy.[13] Our analysis strongly suggests that corporate actors and others with vested interests in the hegemonic model will resist radical, transformative change and the realignment of power relations that would follow in its wake. In the resulting political economic stalemate, the global food system would continue on its unsustainable, inequitable course, assailed by the increasingly devastating impacts of 'extreme climate events' as global warming intensifies and exceeds the limits of the Paris Agreement.

Postscript

Transnational corporate power relations underlie the long-standing geopolitical inequalities and rigidities in world trade and food policy (Friedmann, 2005; McMichael, 2005). These have been exposed most recently by the renewal of hostilities between Russia and Ukraine in February, 2022. Analysing the ensuing global food price inflation – 'Another perfect storm?' – IPES-Food (2022b) calls out the global food order for abject failures of governance to safeguard the poor in the vulnerable low-income countries of the Global South from global food price shocks, the third in the past 15 years. "Progress on reducing hunger has stagnated since 2014/2015, and went fully into reverse in 2020 as the pandemic drove hunger up by 54%, leaving up to 811 million under-nourished" (IPES-Food, 2022b: 4).

In the Global North, the world food price shock and upsurge in fuel and energy prices have combined to cause a 'cost of living crisis'. The annual rate of inflation in the UK, for example, exceeded 10 per cent in 2022, leading some commentators to declare that the 'golden era of cheap food is over'. The Russia–Ukraine war has exacerbated the downward spiral of food insecurity, food poverty, indebtedness, and social injustice experienced by lower income groups at the height of the COVID-19 pandemic in 2020–2021, and which is likely to be compounded by global economic recession in 2023.

An overwhelming fear is that Western governments are responding to the historic opportunity presented by soaring energy and food prices not by accelerating a transition to sustainable production but by weakening their COP26 commitments to renewables and increasing their reliance on fossil fuels. Such a reversal would create a 'perfect storm' of growing intensity, threatening the habitable future of the planet.

Notes

Chapter 1

1. While acknowledging that industrial food systems are found in other countries and regions beyond Western Atlantic nations, this book uses the Global North and modern agro-food systems as synonymous with North America, the European Union, and the United Kingdom.
2. Based on 2019 data, the share of different components in agro-food system emissions are as follows: farmgate emissions – 43.6 per cent; pre- and post-production – 35.2 per cent, and land use change – 21.2 per cent (FAO, 2021).
3. A number of these accounts are noted in Clapp and Moseley (2020).

Chapter 2

1. The Big Tech companies active in the AgTech space include IBM providing weather information, analytics and cloud infrastructure, Amazon with Amazon Web Services, and Microsoft developing Internet of Things capacity (Day, 2019).
2. John Deere launched its first commercially available, fully autonomous or driverless tractor, a version of its 8R row crop tractor, in January, 2022. This announcement followed the US$250 million takeover of Bear Flag, a start-up specialising in the automation of farm machinery (Ellis, 2022a).
3. Sonka and Cheng (2015: 206) warn against the conflation of these concepts, stressing that this powerful analytic capacity is the "differentiating feature of big data", and not "a new buzzword" for the PA technologies that generate the data.
4. In their search for greater conceptual clarity, Kitchin and McArdle (2016) find multiple forms of Big Data and identify velocity and "exhaustivity" as the key boundary markers.
5. The ETC Group (2018b: 10–11) provides a succinct summary of the reach of digital farm service platforms: "Every tool of the data platform impacts every segment of the industrial food chain … every part of the chain uses remote and built-in sensors to gather data, clouds to store data, artificial intelligence (AI) to analyse information, algorithms to manipulate it, and blockchains to distribute it".
6. Philip Howard (2021) notes that John Deere accounts for 80 per cent of North American sales of heavy tractors and combines, and 18 per cent of the global market in farm equipment.
7. For discussion of the vertical integration and cross-sector strategies of leading farm equipment corporations, such as AGCO, CNH, and the Japanese company, Kubata, see the ETC Group (2018).
8. These earlier periods of consolidation are examined at greater length in Chapter 5.
9. This paragraph draws heavily on a case study by the Harvard Business School (2015).
10. Monsanto had previously taken over several companies specialising in farm software tools, notably Precision Planting, before its 2013 acquisition of the Climate Corporation, and bought three more platform companies in 2014 – two specialising in web apps (Climate

Basic, and Climate Pro) and FieldScripts, a planting prescription service (ETC Group, 2018). In turn, the Climate Corporation had earlier pursued its own M&A strategy and continued on this path after its takeover, acquiring Solum, a soil testing company, and 640 Labs, an agro-biotechnology start-up, both in 2014.

[11] This observation is disingenuous, since Monsanto had extensive accumulations of seed and crop protection data and was deeply engaged in data science before its acquisition of Climate Corporation. As Carolan (2017b: 139) notes, "After all, biotechnology is also Big Data. Locating the genes for favourable, and profitable, traits in plants in order to create new seed varieties means sifting through the billions of base pairs in a genome."

[12] This is a reference to the spectacular stock market debut (IPO) of Netscape and its Web browser, Navigator, in 1995, which some say brought the venture capital culture of Silicon Valley to national attention and triggered the dot.com boom of the late 1990s.

[13] According to a recent report (Jia, 2020) ChemChina is merging with its fellow state-owned enterprise, Sinochina. These two enterprises had previously merged their agricultural chemical interests, which now form part of the Syngenta Group.

[14] Bayer has forged cross-sector alliances with the leading farm equipment companies and its M&A and joint venture strategies have targeted firms developing technologies for microbial seed and soil treatments and crop nutrients, which potentially threaten to disrupt fertilizer markets.

Chapter 3

[1] Secretary Perdue's remarks were made at a dairy expo in Wisconsin, where many dairy farmers are in difficulties, and are reported in the Minneapolis Star Tribune of 4 October, 2018 under the editorial headline "Go big or just go". The same phrase, "Get big or get out!" is also attributed to Earl Butz, U.S. Secretary of Agriculture between 1971 and 1976, by Thompson (1998), cited in Gibson and Gray (2019).

[2] The IATP (2020: 15) refer to "Get big or get out" as an "explicit drive in farm policy ... In reality, the policy of discarding smaller scale farmers and promoting industrial scale agriculture has increased rural poverty, frayed the fabric of rural communities, and devastated local environments and waterways".

[3] As MacDonald et al (2018: 2) point out, large family farms "embody a range of distinctive organisational strategies and business practices", including lease and rental agreements, the use of custom service providers and labour contractors, and ownership and operation of multiple farms.

[4] These census returns exaggerate this decline since farms keeping milk cows for their own consumption are included, whereas a state government license is required to sell milk commercially. In 2017, the difference between these two figures was 14,400 farms (MacDonald et al, 2020).

[5] "Today, the largest dairy farms in the country milk over 25,000 cows, usually organised into a series of pods comprised of cow barns or lots, manure storage units, feed bunkers and milking facilities" (MacDonald et al, 2020: 11).

[6] Pope and Sonka (2020) have undertaken a small pilot study of Midwestern farmers using PA technologies in corn and soybean production, who they characterise as "early adopters". Their results indicate that if adoption is considered over a period of years, these technologies make a cumulatively significant contribution to farm profits in all farm size categories, although farms with over 5,000 acres enjoyed the greatest benefits.

[7] This survey is based on a purposive sampling scheme stratified by farm size and major cropping system undertaken in France, Germany, Greece, the Netherlands, Serbia, Spain, and the United Kingdom. Interviews were conducted with farmers (n = 287), evenly divided between adopters and non-adopters of SFT, and agricultural experts (n = 22). For our purposes here, PA and smart farming can be treated as equivalents.

NOTES

8   For the EU, see Morgan et al (2008) and Lamine (2020).
9   For a critical review of a recent initiative by the Federal Communications Commission to bridge the digital divide by extending broadband into poorly served regions of rural America, see Bloch (2020). An estimated 10 million rural inhabitants either have no internet service or lack high-speed access.
10 The authors conducted 78 semi-structured interviews and five focus groups with arable, dairy, and livestock farmers and their professional advisors in the regions of East Anglia, Central, and North Wales, and the counties of Sussex and Devon.
11 In April, 2022, the National Farmers Union, state farmers unions and advocacy organisations directly challenged John Deere's right to repair policy by filing a formal complaint with the US Federal Trade Commission (Hirsch, 2022).
12 For a conceptual and empirical analysis of Farm Hack and interviews with members in the US and Western Europe, see Carolan (2017a, 2017b).
13 In the commercial digital world, corporate open source platforms are profitmaking, which prompts Stallman (2018) to insist that a clear distinction be drawn between open source software and free software. For further discussion, see Rotz et al (2019a).
14 Publications of the socialist science movement include *Science for the People* in the US and *Science for People* in the UK. Other UK sources are the *Radical Science Journal* and occasional papers in the *Journal of the Conference of Socialist Economists (CSE)*.
15 For example, Rotz et al (2019b) analyse how automation is accentuating the unequal, binary structure of agricultural labour markets in North America as robots replace low-skilled jobs and increase opportunities for more highly skilled workers. For example, robotic milking machines take over a repetitive, low-skilled job, often performed by foreign migrant labour, and increase demand for skilled workers in animal health and welfare activities.
16 According to Rotz et al (2019a 212), such knowledge "cannot be included in the algorithmic calculations of (corporate) software. Without being able to integrate such tacit, contextually-specific information, many farmers may struggle to trust or see the value in the outputs of digital analytical tools."
17 More generally, Larry Busch (2019: 102) suggests that Big Data-enabled technologies are introducing a "minute division of labour", which he characterises as the "new Taylorism" or new era of "scientific management" (Taylor, 1911), by extending auditing practices, performance monitoring, and other management control mechanisms, to agro-food supply chains, food processing, food services, and retail activities.

Chapter 4

1   Although currently less prominent, edible insect protein products are being developed by a number of start-ups, including Exo, Chirps, and Enterra. As we note in the text further on, Archer-Daniels Midland (ADM), a major feed processor, has formed a partnership with the French firm, InnovaFeed, to produce insect protein for animal feed.
2   The efficiency and versatility of recombinant-DNA methods were keenly debated in the early 1980s, with some predicting that protein engineering would make it possible to design novel enzymes for specific industrial processes and feed stocks (Ulmer, 1983), whereas others thought that more progress would be made by changing the structure and function of existing enzymes (Perutz, 1985).
3   Although not discussed here, start-ups, such as Good Catch and Ocean Hugger Foods are producing plant-based seafood products, and Finless Food and others are developing cell-cultured fish products.
4   In 2015, Monde Nissin, based in the Philippines, acquired Quorn for US$831 million from the founding company, Marlow Foods, as part of its global diversification strategy. The revival of commercial interest in mycoprotein is also demonstrated by the Better Food

5   Co.'s recent investment in a 13,000 square foot fermentation plant and its introduction of a new line of mycoprotein meat analog ingredients (Poinski, 2021c).
5   Commenting on the intervening three decades before its takeover by Monde Nissin, Sexton (2014: 21) notes that Quorn was being grown on imported corn starch: "What … had begun as a philanthropic and sustainably-minded project … [to] 'feed the world' … had instead produced a specialist and carbon-intensive food product for middle-class, western consumers".
6   According to Poinski (2019), EAT JUST successfully applied for a technical patent for its method of scanning and identifying useful plant proteins, and holds technical patents for its plant-based egg substitutes, Just Mayo and Just Egg.
7   Beyond Meat's eponymous burger is produced from some 15 ingredients, including peas and mung beans, fats, coconut oil, cacao butter, and vegetable oil, with carbohydrates as emulsifiers.
8   The umbrella term 'cellular agriculture' can be confusing, and here we have chosen to follow the usage adopted by Stephens et al (2018). Other researchers, such as Sexton (2018: 2), for example, prefer to distinguish between "acellular" methods, such as the genetic modification of yeast cells to produce milk and egg proteins, and "cellular" methods, including tissue engineering, to cultivate meat from animal-derived stem cells.
9   Technically, cell-cultured heme is not regarded as a genetically modified organism (GMO) and so, despite its use in the production process, the Impossible Burger itself is considered to be free of genetically modified material for regulatory purposes (Froggatt and Wellesley, 2019). According to Brown (2019), soy leghemoglobin was approved by the FDA as a colour additive.
10 These online sources include: The Counter, previously The New Food Economy, Civil Eats, FoodDive, The Spoon, AgFunder News, and Biofuels Digest.
11 For example, in 2019, in order to scale up production and improve distribution, Impossible Foods reached a partnership agreement with the large-scale, national meat supplier, OSI Group.
12 To define synthetic biology, Davies (2018: 2) distinguishes between analytic biology, which focuses on "how naturally evolved living things work", and synthetic biology, which "is concerned with the creation of new living things by deliberate design." Synthetic biology can also modify "existing organisms to do entirely new things" (Davies, 2018: 3), such as making "new biological parts that can be inserted into algae or yeast" (ETC Group, 2018: 20), and used in fermentation processes to produce animal-free protein, for example.
13 Examples of corporate venture capital arms include Tyson New Ventures, Tate & Lyle Ventures, Kellog's Eighteen94Capital, Campbell Soup's Acre Venture Partners, Unilever Ventures, Nestle Venture Capital Fund, and Kraft-Heinz's Evolv. For further details on Big Food's venture capital funds and their investments, see Coyne (2020a).
14 For a comprehensive survey of the investments by major meat corporations in plant-based and cellular meat companies, see Coyne, (2020b). The online source, FoodDive, regularly tracks the activities of individual firms in the plant- and cell-based 'spaces'.
15 This is a joint estimate by the Plant-Based Food Association and the Good Food Institute. Overall, the US plant-based foods market was US$7 billion in 2020, with plant-based milk products accounting for US$4.4 billion, equivalent to 14 per cent of the entire milk market. See GFI (2021) and Cooper (2020) for further details.
16 This sale proved to be controversial, however, as the consortium involved included the private equity firm, Blackstone, which is led by a prominent Trump donor and has been linked to deforestation in the Brazilian Amazon.
17 In response to the discursive offensive by the AP industry, the conventional livestock sector has articulated counter-narratives and mounted a moral challenge of its own by appealing for the conservation of the socio-cultural values embedded in place, wholefoods, artisan

producers, provenance, and rural landscapes, including claims that regenerative agriculture mitigates GHG emissions (Santo and Clinton, 2017).

18   'Feeding the 9 billion by 2050' is a particularly egregious 'promissory narrative', since, as A.K. Sen and others explained many years ago, the problem of global food insecurity and under-nutrition is not the lack of aggregate supply but one of access and distribution. This discourse repeats what Hugh Campbell (2012) calls the "core deceit" of post-Second World War international agricultural policy that global hunger would be resolved by food production and exports from the Global North, a neo-colonial premise that continues to inspire today's food sovereignty movements and peasant organisations.

19   Exceptionally, Impossible Foods has delayed its IPO and continues to rely on private sources, raising US$800 million in 2021 to reach a total overall of US$2 billion.

20   For Scoones (2022), such a nuanced approach is needed to assess the impact of different livestock systems on climate change, biodiversity, livelihoods, landscape conservation, and carbon sequestration, for example.

21   Lynch and Pierrehubert (2019) are critical of LCAs that adopt metrics that reduce all GHGs to CO_2 equivalents and ignore variations in the atmospheric lifespan of different GHGs. For example, methane from the enteric digestion of ruminants is a more powerful GHG than CO_2 but shorter lived in the atmosphere. Santo et al (2020) also qualify comparative estimates of land use efficiencies in these studies by noting that ruminants can graze land unsuitable for crops. See also Scoones (2022).

22   This historical parallel is clear from the comment by Sexton et al (2019: 54) that "AP products join the recent nutricentric trend (Scrinis, 2008) that has seen protein treated as a food category in its own right". In short, the AP industry focuses on a single macronutrient.

23   In an earlier review, Bohrer (2019) concluded that popular meat analogue products are "ultra-processed" but successfully simulate the nutritional specifications of animal meat.

24   Although not a comprehensive list, contributions on this theme include Yoxen (1981); Goodman et al (1987); Kloppenburg Jr. (1988); Boyd, (2001); Boyd et al (2001); Schurman and Kelso (2003); and ETC Group, (2018).

Chapter 5

1   The US Supreme Court ruling in the *Chakrabarty vs. Diamond* case in 1980 allowed life forms to be patented for the first time.

2   The development of molecular biology and the role of structuralist, biochemical, and informationist approaches to genetics are discussed by Allen (1978). See also Kay (1993) and Keller (1995).

3   Schimmelpfennig and King (2006) analysed the agricultural biotechnology patents issued between 1976 and 2000, classified by their original patent holders and their owners in 2002. This revealed that 95 per cent of the patents originally held by seed companies or small agbiotech firms had been acquired by large chemical or multinational corporations. Chemical companies retained all 651 patents for which they were the original owners, but had also acquired 219 patents from agbiotech firms and 451 patents from seed companies (cited by Dunwell, 2012).

4   For discussion of Material Transfer Agreements (MTAs), see Rodriguez (2008).

5   Bt, or *Bacillus thuringiensis*, a common soil bacterium, produces a protein that is toxic to many plant pests.

6   The EPA imposed mandatory Bt monitoring requirements but similar plans for glyphosate monitoring were opposed by Monsanto and allied interest groups (Benbrook, 2018).

7   Benbrook (2018: 388) argues that "The emergence and spread of resistant cotton and corn insects have markedly reduced the value of the Bt-transgenic traits and, in the absence of changes in policy and practice, will eventually render them obsolete, including those

8. expressing multiple Bt toxins". However, these changes are "unlikely given the current political climate in the US" (Benbrook, 2018: 388).
9. This pest-management trajectory continues today. For example, in May 2020, Monsanto–Bayer applied to the USDA for the deregulation of a new GM-corn variety resistant to dicamba, 2, 4-D, glyphosate, glufosinate, and quizalofop.
10. Western Europe is defined as France, Germany, the Netherlands, Belgium-Luxembourg, and Switzerland. The analysis covers the period between 1961, when the FAO began collecting yield data, and 2010.
11. Broadly similar conclusions on the failure of GM crops to raise intrinsic crop yields when compared with conventional crop breeding are reached by Gurian-Sherman (2009).
12. Citing a paper, 'CRISPR under control: small molecule inhibitors of Cas9 identified', *Genetic Engineering News* (3 May, 2019) reports on the identification of small "anti-CRISPR molecules that inhibit the functioning of the Cas9 enzyme and make gene-editing hard to control", adding "As some scientists have known for some time."
13. Archibald (2018: 27) predicts that it is "safe to assume that it will soon cost as much to store DNA sequence information as to generate it."
14. The accelerating pace and falling costs of sequencing technologies are demonstrated by a report that bio-molecular research scientists at the University of California, Santa Cruz have developed a nanopore sequencing technology and algorithm to assemble individual complete human genomes … in about 6 hours and at a cost of about $70 (*Genetic and Biotechnology News*, 5 May, 2020).
15. ZFNs – zinc-finger nucleases; TALEN – transcription activator-like effector nuclease; CRISPR – clustered regularly interspaced short palindromic repeats.
16. Several years ago, there were reports from international projects in Africa that conventional methods, reinforced by marker-assisted breeding, were proving to be more successful in producing drought-tolerant varieties than parallel efforts using transgenic approaches (Gilbert, 2014).
17. One of the complexities of this challenge to mitigate abiotic stress is the problem of pleiotropy, where a gene edit has unintended effects on the phenotypic expression of other genes that negatively affect desirable plant traits. A further difficulty is that a gene edit for drought tolerance, say, may only be effective under certain drought conditions but not in all water-scarce environments. In such cases, new plant traits that are not effective across diverse cropping systems and agro-ecological environments may fail to satisfy corporate investment criteria. I am grateful to Dr Doug Gurian-Sherman for discussing these points with me. The usual disclaimers apply.
18. Oxitec had previously undertaken the experimental release of genetically modified (GM) mosquitoes in Brazil. Following its approval of Experimental Use Permits in Florida and Texas, the EPA in 2022 sanctioned the open environmental release of 2 billion GM mosquitoes in 12 counties of California, the largest such release in the world to date. These decisions have been taken without public scrutiny and debate, least of all in the communities most directly involved. In the case of GM releases in Florida and Texas, see Kofler and Kuzma (2020).
19. Recent research on remediation strategies includes Marshall and Akbari (2018), Medina (2018), and Bulger and Silva (2020).
20. On the risks of self-propagating gene drives, see Esvelt and Gemmell (2017).
21. For varied suggestions to introduce new, far-reaching approaches to the governance, oversight and regulation of gene drives, see Marshall (2010); NASEM (2016); CAST (2018); Convention on Biodiversity (2018); ETC (2018a); Jasanoff and Hurlbut (2018); Medina (2018); Shukla-Jones et al (2018); and Montenegro de Wit (2019, 2020a).
22. On the regulatory path taken by the biotechnology firm, Oxitec, in the case of GM mosquito releases in Florida, see Waltz (2021) and White (2021).

NOTES

22. Countries that had allowed both forms of protection to co-exist before joining UPOV, such as the US, were exempt from this provision.
23. Boyd (2003) cites estimates that Stanford University and the University of California, San Francisco, earned almost US$200 million in royalties by licensing the Cohen-Bayer patent.
24. Mandatory safety precautions were introduced somewhat later in the UK and implemented by an expert committee, the Genetic Manipulation Advisory Group (GMAG).
25. The Animal and Plant Health Inspection Service of the US Department of Agriculture.
26. Provisions for the statutory labelling of GM foods were included in the Genetically Engineered Food Right-to-Know Act, which was introduced in 1999 but failed to garner sufficient Congressional support (Guthman, 2003), and there have since been several other legislative initiatives.
27. In cases where genetic engineering is used during development, it is claimed that the final product does not need to be regulated as a GMO if it no longer contains GM material.
28. On 29 April, 2021, Reuters reported that the European Commission is to review its legislative approach to GMOs, including the regulation of gene edited plants.
29. SECURE: sustainable, ecological, consistent, uniform, responsible, efficient.

Chapter 6

1. The phrase in the chapter heading, between physical space and digital space, is based on an article by Andy Beckett: "Will we ever recover from the great British shopping crisis?" published in *The Guardian*, 12 March, 2021.
2. In this respect, Howard (2016: 20) notes that "for fifteen EU countries, the top five firms controlled a weighted average of 49 per cent of the market in 1999 (Dobson, Waterson and Davies, 2003)".
3. Getir raised US$350 million in June, 2021, giving it a valuation of US$7.5 billion, exceeding the market value of established UK retailers, such as Marks & Spencer and Morrisons (Butler, 2021c). Other online grocery delivery start-ups in the US, Europe, and China also reported impressive post-pandemic funding rounds in 2020–2021, with Xingsheng Youxuan raising the huge sum of US$2 billion, Instacart US$264 million, and Rohlik in the Czech Republic US$320 million (Albrecht, 2021a).
4. Amazon's global e-commerce logistics system reputedly is supported by around 300 fulfilment centres.
5. Weldon (2021) notes that Instacart "provides e-commerce for partner stores, including branded websites and apps with integrated Instacart delivery … This facet of Instacart's business highlights the important point that it is a tech company (not a grocer)".
6. According to Kelloway (2020b), "Grubhub, for instance, has been criticized for creating unique (telephone) numbers for restaurants and charging a $9 commission on all calls, even if they don't result in orders."
7. Bradshaw (2019) describes these firms as "essentially property ventures" that establish and lease out shared kitchen space to third parties.
8. At this writing, Grab is negotiating a 'reverse' merger with the special purpose acquisition company (SPAC), Altimeter Capital Management, for a listing on the Nasdaq Stock Exchange at a valuation of nearly US$40 billion (Lachapelle, 2021).
9. The acquisition of Grubhub has since been described by disgruntled leading investors in TakeawayJustEat as financial mismanagement that has destroyed equity capital as its market value has plunged from US$17.4 billion in early 2021 to US$5.5 billion in early May, 2022. Takeaway reportedly is considering selling Grubhub or finding a 'strategic partner' (Jolly, 2022). In this event, Amazon may well be a possible suitor as it acquired a small equity stake in Grubhub in July, 2022.
10. It later emerged that one of the firm's underwriters, Goldman Sachs, spent £75 million buying Deliveroo shares in an effort to stabilise the price in the first days of trading.

11. In late July, 2021, Deliveroo announced that it will withdraw from the Spanish market following a government decision to introduce legislation to treat 'gig' workers as employees rather than independent contractors (Jolly, 2021).
12. Butler (2021a) reports that the takeaway group Just Eat is now offering the minimum wage, sick pay, and holiday pay to its delivery riders in London, Birmingham, and Liverpool.
13. As Benner et al (2020) note, Proposition 22 "sets a dangerous precedent of enshrining a new substandard category of employment into state law, underscoring the importance of federal legislation against misclassification".
14. For a 'progress report' on direct forms of social action to re-classify gig economy workers as employees, see Held (2022). At issue is a 2019 ruling by the National Labor Relations Board that gig workers are not covered by the Federal law that confers the right to unionise.
15. The socio-technical opportunities created by COVID-19 and its variants are demonstrated by the sales and earnings bonanza enjoyed by the US tech giants (Rushe, 2021).
16. According to a World Food Programme (WFP) press release on 21 April, 2020, projections showed that the number of people in middle- and low-income countries suffering from acute hunger could more than double by the end of the year, rising by 135 million people due to the economic impact of COVID-19 (WFP, 2020).
17. This more comprehensive scheme will complement Tesco's own rapid delivery service, Whoosh, which operates from about 60 of its smaller Tesco Express convenience stores (Butler, 2021d). Other UK supermarkets, including Waitrose, Sainsbury's, and Morrisons, turned to Deliveroo to meet the surge in demand for online home deliveries at the height of the pandemic (Butler, 2022).
18. The prospect of further partnerships along these lines may have been a factor in the recent takeover of Weezy by Getir, as well as stimulating further consolidation in the on-demand, ultra-fast grocery delivery sector (Partridge, 2021b).

Chapter 7

1. This issue has moved toward centre-stage following the sweeping Executive Order on Competition issued on 9 July, 2021 by President Biden, which refers explicitly to consolidation in the agro-food system and the cost-price squeeze on farmers (Kelloway, 2021b).
2. Although pre-tax losses have grown considerably, reports of TakeawayJustEat and Deliveroo for January–June, 2021, indicate that the UK home food delivery sector has continued its rapid sales growth despite the relaxation of lockdown measures and the re-opening of dine-in restaurants (Ambrose, 2021; Partridge, 2021a).
3. The third part of the IPCC's Sixth Assessment Report on ways to reduce human impacts on the climate strongly endorses the significance of a dietary change to plant-based protein in the world's richest countries and among the wealthy globally (Harvey and Tremlett, 2021).
4. These contributions include the CFS/HLPE reports, *Food Security and Nutrition: Building a Global Narrative Toward 2030* (2020), and *Agro-Ecological and Other Innovative Approaches to Sustainable Agriculture and Food Systems that Enhance Food Security and Nutrition* (2021), IPES-Food's *Too Big to Feed: Exploring the Impact of Mega-Mergers, Consolidation and Concentration of Power in the Agri-Food Sector* (2017), and IPES-Food and ETC Group's *A Long Food Movement: Transforming Food Systems by 2045* (2021).
5. At the request of the CFS, the HLPE, in conjunction with IPES-Food and other organisations, recently elaborated "A unifying framework for food systems transformation. A call for governments, private companies and civil society to adopt 13 key principles" (July, 2021). At this writing, negotiations to accept these principles, or even weaker versions, are deadlocked (Heffron and Varghese, 2021).

NOTES

6. Professor Boyd Swinburn is a co-author of *The Global Syndemic: Obesity, Under-nutrition and Climate Change* (Swinburn et al, 2019), and served as co-chair of the Commission.
7. AGRA is heavily funded by the Bill and Melinda Gates Foundation.
8. Opponents of the proposed science–policy interface pointed out that the HLPE of the UN Commission on World Food Security already fulfils this role and argued that the UNFSS should build on and strengthen this existing institutional foundation.
9. Disputes provoked by the UNFSS planning process led to open letters of protest to governments and policymakers, as well as numerous calls to boycott proceedings. A critical point in this wave of protest occurred on 26 July, 2021, when the IPES-Food panel and its members withdrew from preparations for the UNFSS. See IPES-Food (2021b).
10. Under pressure, the UNFSS recognised agroecology but continued to ignore its socio-cultural dimensions, 'bundling' it with other technologies, such as PA, Big Data analytics, biotechnology and so on. For further details, see Montenegro de Wit et al (2021b).
11. For recent discussion of the polarised debate between technocratic and eco-centric approaches, see, for example, Levidow et al (2019) and Kneafsey et al (2021).
12. Scientists are predicting with growing confidence that global warming will exceed the Paris Agreement target level of 1.5 °C in the next 5 years (Carrington, 2022).
13. In the case of the US, see Cochrane (1979), for Western Europe, Allaire and Boyer 1995), and Goodman and Redclift (1991) for the UK.

References

AgFunder News (2019) 'Syngenta CEO: "climate change is our biggest challenge"', AgFunder News, [online] 14 May, Available from: https://agfundernews.com/syngenta-ceo-climate-change-will-be-our-biggest-challenge

Ainsworth, C. (2015) 'Agriculture: A new breed of edits', *Nature*, 528: 15–16.

Albrecht, C. (2020a) 'Be like Walmart and swing for the geofences', The Spoon, [online] 19 May, Available from: https://thespoon.tech/be-like-walmart-and-swing-for-the-geofences/

Albrecht, C. (2020b) 'SKS 2020: Impossible Foods CEO on cell-based meat: "It's never going to be a thing"', The Spoon, [online] 14 October, Available from: https://thespoon.tech/sks-2020-impossible-foods-ceo-on-cell-based-meat-its-never-going-to-be-a-thing/

Albrecht, C. (2021a) 'Will grocery delivery become the next utility?' The Spoon, [online] 26 March, Available from: https://thespoon.tech/will-grocery-delivery-become-the-next-utility/

Albrecht, C. (2021b) 'Self-driving delivery speeds up', The Spoon, [online] 16 April, Available from: https://thespoon.tech/self-driving-delivery-speeds-up/

Albrecht, C. (2021c) 'Why e-commerce 10-minute grocery start-ups are not the next Kozmo.com', The Spoon, [online] 5 June, Available from: https://thespoon.tech/why-10-minute-grocery-startups-are-not-the-next-kozmo-com/

Albrecht, C. (2021d) 'Food Rocket's CEO about the nuts and bolts of speedy grocery delivery', The Spoon, [online] 20 July, Available from: https://thespoon.tech/food-rockets-ceo-about-the-nuts-and-bolts-of-speedy-grocery-delivery/

Albrecht, C. (2021e) 'Retailers: don't fret over online grocery's downward trend', The Spoon, [online] 27 July, Available from: https://thespoon.tech/retailers-dont-fret-over-online-grocerys-downward-trend/

Albrecht, C. (2021f) 'Let's unpack the possible DoorDash+ Gorillas deal', *The Spoon Newsletter*, [online] 8 August, Available from: https://thespoon.tech/lets-unpack-the-possible-doordash-gorillas-deal/

Allaire, G. and Boyer, R. (eds) (1995) *La grande transformation de l'agriculture. Lectures conventionalistes et regulationnistes*, Paris: INRA/ECONOMICA.

Allen, G.E. (1978) *Life Science in the Twentieth Century*, Cambridge: Cambridge University Press.

REFERENCES

Alphey, L.S., Cristani, A., Randazzo, F., and Akbari, O.S. (2020) 'Opinion: standardizing the definition of gene drive', *Proceedings of the National Academy of Sciences*, 117(49): 30864–7.

Ambrose, J. (2021) 'Deliveroo surges in pandemic as orders double', The Guardian, 12 August.

Anderson, B. (2010) 'Pre-emption, precaution and preparedness: anticipatory action and future politics', *Progress in Human Geography*, 34(6): 777–8.

Archibald, J. (2018) *Genomics: A Very Short Introduction*, Oxford: Oxford University Press.

Arthur, W.B. (1989) 'Competing technologies, increasing returns and lock-in by historical events', *Economic Journal*, 99: 116–31.

Ash, J., Kitchin, R., and Leszczynski, A. (2016) 'Digital turn, digital geographies?', *Progress in Human Geography*, 42(1): 25–43.

Ash, J., Kitchin, R. and Leszczynski, A. (Eds.) (2019) *Digital Geographies*, London: Sage.

Askew, K. (2021a) 'JBS on its acquisition of Vivera: "It is an important step in our ton our plant-based trajectory"', Food Navigator, [online] 20 April, Available from: https://www.foodnavigator.com/Article/2021/04/20/JBS-on-its-acquisition-of-Vivera-It-is-an-important-step-in-our-plant-based-trajectory

Askew, K. (2021b) 'Is food and agriculture missing from the political debate at COP26?' Food Navigator, [online] 4 November, Available from: https://www.foodnavigator.com/Article/2021/11/04/Is-food-and-agriculture-missing-from-the-political-debate-at-COP26

Augere-Granier, M.-L. (2018) 'The EU dairy sector', European Parliamentary Research Service, Briefing PE 630345-December, 2018.

Barwise, P. and Watkins, L. (2018) 'The evolution of digital dominance. How and why we got to GAFA' in M. Moore and D. Tambini (eds) *Digital Dominance: The Power of Google, Amazon, Facebook, and Apple*, New York: Oxford University Press, pp 20–49.

Bellamy Foster, J. and Suwandi, I. (2020) 'COVID-19 and catastrophe capitalism. Commodity chains and epidemiological-economic crises', *Monthly Review*, 72(2): 545–59.

Benbrook, C.M. (2012) 'Impacts of genetically engineered crops on pesticide use in the U.S. – the first sixteen years', *Environmental Sciences Europe*, 24(1): 1–13.

Benbrook, C.M. (2018) 'Why regulators lost track and control of pesticide risks: lessons from the case of glyphosate-based herbicides and genetically-engineered crop technology', *Current Environmental Health Reports*, 5: 387–95.

Bene, C. (2022) 'Why the Great Food Transformation may not happen: a deep-dive into our food system's political economy, controversies and politics of evidence', *World Development*, 154: 1–14.

Benner, C. and Mason, S., with Carre, F., and Tilly, C. (2020) *Delivering Insecurity: E-commerce and the Future of Working in Food Retail*, Berkeley: UC Berkeley Labor Center and Working Partnerships USA, [online], Available from: https://laborcenter.berkeley.edu/delivering-insecurity/

Bijker, W.J. (1995) *Of Bicycles, Bakelites and Bulbs: Towards a Theory of Sociotechnical Change*, Cambridge, MA: MIT Press.

Bloch, S. (2019) 'If farmers sold their data instead of giving it away, would anyone buy?' Counter, [online] 19 July, Available from: https://thecounter.org/farmobile-farm-data/

Bloch, S. (2020) 'FCC announces "biggest and boldest step yet" to expand rural broadband', The Counter, [online] 30 October, Available from: https://thecounter.org/fcc-20-billion-dollar-rural-broadband-auction/

Bloomberg (2019) 'Amazon came for supermarkets and grocers are fighting back', Bloomberg, [online] 3 December, Available from: https://www.bloomberg.com

Bogue, A. G. (1968) *From Prairie to Cornbelt*. Chicago: Quadrangle Books.

Bogue, A. G. (1983) 'Changes in mechanical and plant technology: the cornbelt, 1910–1940', The *Journal of Economic History*, 43(1).

Böhm, S., Spierenberg, M., and Lang, T. (2020) 'Fruits of our labour: work and organisation in the global food system', *Organisation*, 27(2): 195–212.

Bohrer, B.M. (2019) 'An investigation of the formulation of the nutritional composition of modern meat analogue products', *Food Science and Human Wellness*, 8(4): 320–9.

Boyd, W. (2001) 'Making meat: science, technology, and American poultry production', *Technology and Culture*, 42: 631–64.

Boyd, W. (2003) 'Wonderful potencies? Deep structure and the problem of monopoly in agricultural biotechnology', in R.A. Schurman and D.D. Takahashi Kelso (eds) *Engineering Trouble: Biotechnology and its Discontents*, Berkeley, CA: University of California Press, pp 24–62.

Boyd, W., Prudham, S.W., and Schurman, R.A. (2001) 'Industrial dynamics and the problem of nature', *Society and Natural Resources*, 14: 555–70.

Bradshaw, T. (2014) 'Food 2.0 – the future of what we eat', Financial Times, 31 March.

Bradshaw, T. (2019) 'The start-ups building "dark kitchens" for Uber Eats and Deliveroo', Financial Times, 21 May.

Bronson, K. (2018) 'Smart farming: including rights holders for responsible innovation', *Technology Innovation Management Review*, 8(2): 7–14.

Bronson, K. (2019) 'Looking through a responsible technology lens at uneven engagements with digital farming', *NJLS – Wageningen Journal of Life Sciences*, 90–91(1): 1–6, DOI: 10.1016/j.njas.2019.03.001.

Bronson, K. and Knezevic, I. (2016) 'Big Data in food and agriculture', *Big Data and Society*, January-June: 1–5.

Brown, H.C. (2019) 'After new plant blood gets FDA approval, the Impossible Burger is set to hit supermarket shelves', The New Food Economy, [online] 31 July, Available from: https://thecounter.org/plant-blood-heme-fda-approval-impossible-burger/

Bruce, A. and Spinardi, G. (2018) 'On a wing and a prayer: eco-modernisation, epistemic lock-in, and the barriers to greening aviation and ruminant farming', *Energy Research and Social Science*, 40(June): 36–44.

Bud, R. (1994) *The Uses of Technology: A History of Biotechnology*, Cambridge: Cambridge University Press.

Bulger, E. and Silva, A. (2020) 'Gene drive control worry eased by genetic neutralising elements', Genetic Engineering and Biotechnology News, [online] 21 October, Available from: https://www.genengnews.com/news/gene-drive-control-worry-eased-by-genetic-neutralizing-elements/

Burt, A. (2003) 'Site-specific selfish genes as tools for the control and genetic engineering of natural populations', *Proceedings of the Royal Society B: Biological Sciences*, 27: 921–8.

Burwood-Taylor, L. (2017) 'Global sales of meat substitutes in 2016 reach US$ 4 billion', AgFunder News, [online] 27 February, Available from: https://agfundernews.com

Busch, L. (2019) 'The new autocracy in food and agriculture', in G. Allaire and B. Daviron (eds) *Ecology, Capitalism and the New Agricultural Economy: The Second Great Transformation*, London: Routledge, pp 95–109.

Butler, P. (2021) 'UK charity gives out 2.5m food parcels as need hits historic high', The Guardian, [online] 22 April. Available from: https://www.theguardian.com/society/2021/apr/22/uk-charity-trussell-trust-gives-out-record-25m-food-parcels-to-meet-historic-levels-of-need

Butler, S. (2021a) 'Thousands of Addison Lee staff could receive £10,000 payout over worker rights', The Guardian, [online] 23 April. Available at: https://www.theguardian.com/business/2021/apr/22/addison-lee-drivers-payout-workers-rights-court-appeal

Butler, S. (2021b) 'Fast food: the new wave of delivery services bringing groceries in minutes', The Guardian, [online] 22 May. Available at: https://www.theguardian.com/business/2021/may/22/fast-food-the-new-wave-of-delivery-services-bringing-groceries-in-minutes

Butler, S. (2021c) 'Delivery firm Getir to expand into US after raising $350 million', The Guardian, [online] 5 June. Available at: https://www.theguardian.com/business/2021/jun/04/delivery-firm-getir-to-expand-into-us-after-550m-funding-round

Butler, S. (2021d) 'Tesco and Gorillas join forces to test 10-minute deliveries', The Guardian, [online] 29 October. Available at: https://www.theguardian.com/business/2021/oct/28/tesco-and-gorillas-join-forces-to-test-10-minute-delivery-service

Butler, S. (2022) 'Aldi ends Deliveroo deliveries as customers return to stores', The Guardian, [online] 21 January. Available at: https://www.theguardian.com/business/2022/jan/20/aldi-ends-deliveroo-deliveries-online-groceries-peak

Buttel, F.H. (1983) 'Beyond the family farm', in G.F. Summers (ed) *Technology and Social Change in Rural Areas*, Boulder, CO: Westview Press, chapter 5.

Buttel, F.H. and Newby, H. (eds) (1980) *The Rural Sociology of the Advanced Societies*, Montclair, NJ: Allanheld Osmun.

Byington, L. (2020) 'How two years of changes in dairy led to two major bankruptcies', FoodDive, [online] 4 March, Available from: https://www.fooddive.com/news/how-two-years-of-changes-in-dairy-led-to-two-major-bankruptcies/573283/

Campbell, H. (2012) 'Let us eat cake? Historically reframing the problem of world hunger and its purported solutions' in R. Rosin, P. Stock, and H. Campbell (eds) *Food Systems Failure: The Global Food Crisis and the Future of Agriculture*, New York: Earthscan, pp 30–45.

Canfield, M., Anderson, M.D., and McMichael, P. (2021) 'UN Food Systems Summit: dismantling democracy and resetting corporate control of food systems', *Frontiers in Sustainable Food Systems*, [online] 13 April, DOI: 10.3389/fsufs.2021.661552.

Caplan, P. (2020a) *Food Poverty and Charity in the UK: Food Banks, the Food Industry and the State*, London: Goldsmiths University of London.

Caplan, P. (2020b) 'Struggling for food in a time of crisis: responsibility and paradox', *Anthropology Today*, 36(3): 8–10.

Caraher, M. (2020) 'Struggling for food in a time of crisis: comments on Caplan', *Anthropology Today*, 36(3): 26–27.

Caraher, M. and Furey, S. (2021) 'Debt and diet', Food Research Collaboration [Blog] 22 July, Available from: https://www.foodresearch.org/blogs/debt-and-diet/

Carbonell, I.M. (2016) 'The ethics of big data in big agriculture', *Internet Policy Review*, 5(1), DOI: 1014763/2016.1.405.

Carolan, M. (2017a) 'Agro-food governance and life itself: food politics and the intersection of code and affect', *Sociologia Ruralis*, 57 (S1): 816–35.

Carolan, M. (2017b) 'Publicising food; big data, precision agriculture and co-experimental techniques of addition', *Sociologia Ruralis*, 57(2): 135–54.

Carolan, M. (2018) '"Smart" farming techniques as political ontology: access, sovereignty and the performance of neoliberal and not-so-neoliberal worlds', *Sociologia Ruralis*, 58(4): 745–64.

Carolan, M. (2019) 'Rural sociology revival; engagements, enactments and affectments for uncertain times', *Sociologia Ruralis*, 60(1): 284–302.

Carolan, M. (2020) 'Automated agrifood futures: robotics, labour and the distributed politics of digital agriculture', *Journal of Peasant Studies*, 47(1): 184–207.

Carolan, M. (2022) 'Digitalisation as politics: smart farming through the lens of weak and strong data', *Journal of Rural Studies*, 91: 208–16.

Carrington, D. (2020) '"Most realistic" plant-based steak revealed', The Guardian, [online] 11 January. Available at: https://www.theguardian.com/food/2020/jan/10/most-realistic-plant-based-steak-revealed

Carrington, D. (2021) 'Europe and US could reach "Peak meat" in 2025', The Guardian, [online] 24 March. Available at: https://www.theguardian.com/environment/2021/mar/23/europe-and-us-could-reach-peak-meat-in-2025-report

Carrington, D. (2022) 'Climate limit of 1.5C close to being broken, scientists warn', The Guardian, [online] 10 May. Available at: https://www.theguardian.com/environment/2022/may/09/climate-limit-of-1-5-c-close-to-being-broken-scientists-warn

Carson, R. (1962) *Silent Spring*, New York: Houghton Mifflin.

CAST: Council for Agricultural Science and Technology (2018) 'Genome editing in agriculture: methods, applications and governance', The Need for Agricultural Innovation to Sustainably Feed the World by 2050, Issue Report Number 60, Ames, IA: CAST.

CB Insights (2020) 'The next shopping and delivery battleground: why Amazon, Walmart and smaller retailers are betting on micro-fulfillment', CB Insights: Research Briefs, [online] 21 July, Available from: https://www.cbinsights.com/research/micro-fulfillment-tech-shipping-retail/

Cherry, M. (2016) 'Beyond misclassification: the digital transformation of work', *Comparative Labour Law and Policy*, 37(3): 577–602.

Clapp, J. (2021) 'Explaining growing glyphosate use: the political economy of herbicide-dependent agriculture', *Global Environmental Change*, 67, DOI: 10.1016/j.gloenvcha.2021.102239.

Clapp, J. and Moseley, W. (2020) 'This food crisis is different: COVID-19 and the fragility of the neo-liberal food security order', *Journal of Peasant Studies*, 47(7): 1393–1417.

Clark, M.A., Domingo, N.G.G., Colgan, K., Thakar, S.K., Tilman, D., Lynch, J., Azevedo, I.L., and Hill, J.D. (2020) 'Global food system emissions could preclude achieving the 1.5° and 2°C degrees climate change targets', *Science*, 370(6517): 705–8.

Clinton, P. (2018) 'What's behind Amazon's obsession with food?' The Counter, [online] 24 April, Available from: https://thecounter.org/amazon-a-whole-foods-growth-profit/

Cochrane, W.W. (1979) *The Development of American Agriculture: A Historical Analysis*, Minneapolis, MN: University of Minnesota Press.

Cohen, J. (2020) 'The latest round in the CRISPR patent battle has an apparent victor but the fight continues', *Science*, 1 September, DOI: 10.1126/science.abe7573.

Colin, N. (2021) 'Deliveroo: is it worth it?' Sifted, [online] 31 March, Available from: https://sifted.eu/articles/deliveroo-ipo-valuation/

Convention on Biological Diversity (2018) Decision adopted by the Conference of the Parties to the Convention on Biological Diversity; Synthetic Biology, CBD/COP/DEC/14/19.

Cooper, B. (2020) 'Plant-based food sales top US$5 billion in the US', Just-Food, [online] 4 March, Available from: https://www.just-food.com

Corporate Europe Observatory (2016) 'Biotech lobby's push for new GMOs to escape regulation', Corporate Europe Observatory, [online] 2 February, Available from: https://corporateeurope.org/en/food-and-agriculture/2016/02/biotech-lobby-push-new-gmos-escape-regulation

Corporate Europe Observatory (2018) 'Embracing nature': biotech industry spin seeks to exempt GMOs from regulation', Corporate Europe Observatory, [online] 14 May, Available from: https://corporateurope.org/en/food-and-agriculture/2018/05/embracingnature

Corporate Europe Observatory (2019) 'US pressure on EU to de-regulate new GM'. Corporate Europe Observatory, [online] 24 July, Available from: https://corporateeurope.org/en/2019/07/us-pressure-eu-de-regulate-new-gm

Cosgrove, E. (2017) 'How do farm hackers view venture-backed agtech?' AgFunderNews, [online] 11 December, Available from: https://agfundernews.com/how-do-farm-hackers-view-venture-backed-agtech.html

Cosgrove, E. (2018) 'Why Syngenta acquired FarmShots: "It's definitely a race"'. AgFunderNews, [online] 16 March, Available from: https://agfundernews.com/syngenta-acquired-farmshots

Cotter, J., Kawall, K., and Then, C. (2020) *New Genetic Engineering Technologies: Report of the results from the RAGES project 2016–2019*, [online] Available from: www.testbiotech.org/projekt_rages

Coyle, D. (2018) 'Platform dominance: the shortcomings of anti-trust policy', in M. Moore and D. Tambini (eds) *Digital Dominance: The Power of Google, Amazon, Facebook, and Apple*, New York: Oxford University Press, pp 50–70.

Coyne, A. (2020a) 'Big Food's stake in the future: in-house venture capital funds', Just-Food, [online] 12 February, Available from: https://www.just-food.com/features/big-foods-stake-in-the-future-in-house-venture-capital-funds/

Coyne, A. (2020b) 'Eyeing alternatives – meat companies with stakes in meat-free', Just-Food, [online] 5 March, Available from: https://www.just-food.com/features/eyeing-alternatives-meat-companies-with-stakes-in-meat-free-and-cell-based-meat/

Coyne, A. (2020c) 'Plant-based priorities – dairy companies with stakes in dairy-free', Just-Food, [online] 5 March, Available from: https://www.just-food.com

Cressey, D. (2013) 'Transgenes: a new breed', *Nature*, 497: 27–9.

Cusumano, M., Gawer, A., and Yoffie, D.B. (2019) *The Business of Platforms: Strategy in the Age of Digital Competition, Innovation, and Power*, New York: Harper-Collins.

Dance, A. (2015) 'Core concept: CRISPR gene editing', *Proceedings of the National Academy of Sciences*, 112(24): 6245–6.

Day, S. (2019) 'AgTech landscape 2019: 1,600 plus startups innovating on the farm and in the "messy middle"', AgFunder News, [online] 4 June, Available from: https://agfundernews.com/2019-06-04-agtech-landscape-2019-1600-startups

Davies, J.A. (2018) *Synthetic Biology: A Very Short Introduction*, Oxford: Oxford University Press.

Deliveroo Editions website, [online], Available from: https://restaurants.deliveroo.com/en-gb/editions

DeLonge, M.S., Miles, A., and Carlisle, L. (2016) 'Investing in the transition to sustainable agriculture', *Environmental Science and Policy*, 55(Part 1): 266–73.

Dimitri, C. and Heffland, A. (2020) 'From farming to food systems: the evolution of US agricultural production and policy into the 21st century', *Renewable Agriculture and Food Systems*, 35(4): 391–406.

Dobson, P.W., Waterson, M., and Stephens, S.W. (2003) 'The patterns and implications of increasing concentration in European food retailing', *Journal of Agricultural Economics*, 54(1): 111–15.

Dodge, M. (2019) 'Rural', in J. Ash, R. Kitchin, and A. Leszczynski (eds) *Digital Geographies*, London: Sage, pp 36–48.

Dosi, G. (1982) 'Technological paradigms and technological trajectories: a suggested interpretation of the determinants and directions of technological change', *Research Policy*, 11: 142–62.

Dunhill, P. (1981) 'Biotechnology and industry', *Chemistry and Industry*, 4 April.

Dunwell, J.M. (2011) 'Foresight project on global food and farming futures. Crop biotechnology: prospects and opportunities', *Journal of Agricultural Science*, 149(S1): 17–21.

Dunwell, J.M. (2012) 'Patents for plants: context and current status', *Acta Horticulturae*, 941: 125–38.

Dunwell, J.M. (2014) 'Genetically modified (GM) crops: European and transatlantic divisions', *Molecular Plant Pathology*, 15(2): 119–21.

Egelie, K.J., Graf, G.D., Strand, S.P., and Johansen, B. (2016) 'The emerging landscape of CRISPR-Cas9 gene editing technology', *Nature Biotechnology*, 34(10): 1025–31.

Ellis, J. (2020) 'Syngenta restructures, rebrands and re-launches after taking $5.6 billion Chinese biz on board'. AgFunder News, [online] 18 June, Available from: https://agfundernews.com/syngenta-group-relaunches-after-taking-5-6bn-chinese-biz-on-board

Ellis, J. (2021a) 'Beyond Meat signs global supply deals with McDonald's, KFC and Pizza Hut', AgFunder News, [online] 1 March, Available from: https://agfundernews.com/beyond-meat-signs-global-supply-deals-with-mcdonalds-yum-brands

Ellis, J. (2021b) 'Nestle to enter cell-cultured meat market in partnership with Israeli start-up', AgFunder News, [online] 13 July, Available from: https://agfundernews.com/nestle-to-enter-cell-cultured-protein-market-in-tie-up-with-future-meat

Ellis, J. (2022a) 'Brief: Jonn Deere unveils first commercially available, fully autonomous tractor', AgFunder News, [online] 5 January, Available from: https://agfundernews.com/john-deere-unveils-its-first-commercially-available-fully-autonomous-tractor

Ellis, J. (2022b) 'The lowdown: Impossible Foods sues Motif FoodWorks over infringement of heme patent', Agfunder News, [online] 10 March, Available from: https://agfundernews.com/heme-patent-impossible-foods-sues-motif-foodworks

ENSSER: European Network of Scientists for Social and Environmental Responsibility (2017) *Statement on New Genetic Modification Techniques Should be Strictly Regulated as GMOs*, Berlin: ENSSER.

Esvelt, K.M. and Gemmell, N.J. (2017) 'Conservation demands safe gene drives', *PLOS Biology*, [online] 16 November, DOI: 10.1371/journal.pbio.2003850.

ETC Group (2013) 'Putting the cartel before the horse ... and farm, seeds, soil, peasants: a report on the state of corporate concentration', *ETC Communique*, 111, Ottawa, Ontario: ETC.

ETC Group (2016a) *Software vs. Hardware vs. Nowhere: Briefing Document*, December, Ottawa, Ontario: ETC.

ETC Group (2016b) *Reckless Gene Driving: Genes and the End of Nature. A Briefing from the Civil Working Group on Gene Drives*, Ottawa, Ontario: ETC.

ETC Group (2018a) *Forcing the Farm: How Gene Drive Organisms Could Entrench Industrial Agriculture and Threaten Food Sovereignty*, Ottawa, Ontario: ETC.

ETC Group (2018b) *Blocking the Chain: Industrial Food Chain Concentration, Big Data Platforms and Food Sovereignty Solutions*, Ottawa, Ontario: ETC.

ETC Group (2021a) Hijacking Food Systems: Technofix Takeover at the FSSI, Communique 118, [online] 23 July, Available from: www.etcgroup.org/content/hijacking-food-systems-technofix-takeover-fss

ETC Group (2021b) 'COP26 resembles global trade fair for geoengineeers, Big Ag and Big Data', Press release, [online] 16 November, Available from: www.etcgroup.org/content/cop26-resembles-global-trade-fair-geoengineers-big-ag-and-big-dat

Eurostat (2018) 'Farms and farmland in the European Union – statistics', Eurostat, [online] November, Available from: https://ec.europa.eu/eurostat/statistics-explained/index.php?title=Farms_and_farmland_in_the_European_Union_-_statistics

FAIRR – A Collier Initiative (2019) *Appetite for Disruption. How Leading Food Companies are Responding to the Alt-Protein Boom*, London: FAIRR.

FAO: Food and Agriculture Organisation of the United Nations (2021) 'FAOSTAT Emissions shares', FAO, [online], Available from: https://www.fao.org/faostat/en/#home

Farm Hack (2018) Farm Hack, [online], Available from: https://farmhack.org/tools

Fassler, J. (2018a) 'What the alt-protein revolution tells us about the future of eating', The Counter, [online] 1 March, Available from: https://newfoodeconomy.orgthecounter.org/silicon-valley-just-alt-protein-clean-meat/

Fassler, J. (2018b) 'I visited Amazon's new 4-star store – a glimpse into the big data-enabled future of brick-and-mortar', The Counter, [online] 27 September, Available from: https://thecounter.org/amazon-4-star-customer-data-brick-mortar-retail-future/

Fassler, J. (2021) 'Lab-grown meat is supposed to be inevitable, the science tells a different story', The Counter, [online] 22 September, Available from: https://thecounter.org/lab-grown-cultivated-meat-cost-at-scale/

Feehan, P. (2021) 'Dietary change at COP26: the missing ingredient', Food Research Collaboration, [online] 11 November, Available from: www.foodresearch.org/blogs/dietary-change-at-cop26-the-missing-ingredient/

Fernandez-Cornejo, J., Wechsler, S., Livingston, M., and Mitchell, L. (2014) *Genetically Engineered Crops in the United States*, Economic Research Report Number 162, Washington, DC: U.S. Department of Agriculture, Economic Research Service.

Field Agent Marketing (2021) *The State of Digital Grocery*, Field Agent, [online], Available from: https://blog.fieldagent.net/report-the-state-of-digital-grocery-2021

Friedland, W.H., Busch, L., Buttel, F.H., and Rudy, A.P. (eds) (1991) *Towards a New Political Economy of Agriculture*, Boulder, CO: Westview.

Friedmann, H. (2005) 'From colonialism to green capitalism: social movements and the emergence of food regimes', in F.H. Buttel and P. McMichael (eds) *New Directions in the Sociology of Global Development: Research in Rural Sociology and Development*, Volume11, Amsterdam: Elsevier, pp 227–64.

Froggatt, A. and L. Wellesley (2019) *Meat Analogues: Considerations for the EU*, London: Chatham House.

Galinsky, E. and Hillbeck, A. (2018) 'European Court of Justice ruling regarding new genetic engineering methods scientifically justified: a commentary on the biased reporting about the recent ruling', *Environmental Sciences Europe*, 30(52), DOI 10.1186/s12302-018-0182-9.

GFI: Good Food Institute (2021) *2020 US State of the Industry Report: Plant-based Meat, Eggs and Dairy*, Washington, DC: Good Food Institute.

Gibson, J.W. (2019) 'Automated agriculture: precision technologies, agbots and the fourth industrial revolution', in J.W. Gibson and S.E. Alexander (eds) *In Defense of Farmers: The Future of Agriculture in the Shadow of Corporate Power*, Lincoln, NB: University of Nebraska Press, pp 135–73.

Gibson, J.W. and Gray, B.J. (2019) 'The price of success: land consolidation and the transformation of communities in western Kansas', in J.W. Gibson and S.E. Alexander (eds) *In Defense of Farmers: The Future of Agriculture in the Shadow of Corporate Power*, Lincoln, NB: University of Nebraska Press, pp 325–61.

Gilbert, N. (2014) 'Cross-bred crops get fit faster', *Nature*, 513(7518): 292.

Giri, A.K., Subedi, D., Todd, J.E., Litkowski, C., and Witt, C. (2021) 'Off-farm income a major component of total farm income for most farm households in 2019', AgEcon Search, [online] 7 September, Available from: https://ageconsearch.umn.edu/record/313519/?ln=en

Gloy, B.A. and Widmar, D.A. (2014) 'Long-term trends in farm demographics, debt use, and land ownership: implications for the financial needs of US farming', *A Report for the Farm Credit Coordinating Committee*.

Goodman, D. and Redclift, M.R. (1991) *Refashioning Nature: Food, Ecology and Culture*, London: Routledge.

Goodman, D., Sorj, B., and Wilkinson, J. (1987) *From Farming to Biotechnology: A Theory of Agro-industrial Development*, Oxford: Basil Blackwell.

Gurian-Sherman, D. (2009) *Failure to Yield: Evaluating the Performance of Genetically Engineered Crops*, Cambridge, MA: Union of Concerned Scientists (UCS).

Guthman, J. (2003) 'Eating risk: the politics of labelling genetically modified foods' in R.A. Schurman and D. Kelso (eds.) *Engineering Trouble: Biotechnology and Its Discontents*, Berkeley, CA.: University of California Press, pp. 130–151

Haerlin, B. (2020) 'The making of a paradigm shift', in H.R. Herren, B. Haerlin and IAASTD+10 Advisory Group (eds) *Transformation of Our Food Systems: The Making of a Paradigm Shift*, Berlin: Zukenftsstiftung Landwirtschaft and the Biovision Foundation, pp 17–20.

Hames, S. (2020) 'Homegrown relief: farming communities tackle the rise in suicides' *Christian Science Monitor*, 25 September.

Harker, R.N., Donovan, J.Y., Blackshaw, R.E., Beckie, H.J., Malloer-Smith, C., and Maxwell, B.D. (2012) 'Our view', *Weed Science*, 60: 143–4.

Harvard Business School (2015) 'Monsanto and Climate Corp: Big Data transforming the agriculture industry', https://d3.harvard.edu/platform-digit/submission/monsanto-climate-corp-big-data-transforming-the-agriculture-industry/#

Harvey, F. and Tremlett, G. (2021) 'Rich must change life-styles – climate report leak', The Guardian, 13 August.

Heffernan, W.D., Hendrickson, M., and Gronski, R. (1999) *Consolidation in the Food and Agriculture System: Report to the National Farmers Union*, Columbia, MO: Department of Sociology, University of Missouri.

Heffron, C. and Varghese, S. (2021) 'Q & A: The United Nations agroecology negotiations and food system summit', IATP, [online] 23 July, Available from: https://www.iatp.org/un-agroecology-negotiations-food-systems-summit

Heinemann, J. (2020) 'Assessment of modern biotechnologies', in H.R. Herren, B. Haerlin and IAASTD+10 Advisory Group (eds) *Transformation of Our Food Systems: The Making of a Paradigm Shift*, Berlin: Zukenftsstiftung Landwirtschaft and the Biovision Foundation, pp 111–15.

Heinemann, J., Massaro, M., Coray, D.S., Agapito-Tenfen, S. and Wen, J.D. (2014) 'Sustainability and innovation in staple crop production in the US Midwest', *International Journal of Agricultural Sustainability*, 12(1): 71–88.

Held, L. (2022) 'The next frontier of labor organizing: food delivery workers', Civil Eats, [online] 4 May, Available from: https://civileats.com/2022/05/04/the-next-frontier-of-labor-organizing-food-delivery-workers/

Hendrickson, M.K. (2020) 'COVID lays bare the brittleness of a centralised and concentrated food system', *Agriculture and Human Values*, 37: 579–80.

Hendrickson, M.K., Howard, P.M., and Constance, D.H. (2019) 'Power, food, and agriculture: implications for farmers, consumers and communities', in J.W. Gibson and S.E. Alexander (eds) *In Defense of Farmers: The Future of Agriculture in the Shadow of Corporate Power*, Lincoln, NB: University of Nebraska Press, pp 13–61.

Hendrickson, M.K., Howard, P.M., Miller, E.M., and D.H. Constance (2020) *The Food System: Concentration and its Impacts. A Special Report for the Family Farm Action Alliance*, 19 November.

Herren, H.R. (2020) 'Introduction', in H.R. Herren, B. Haerlin and IAASTD+10 Advisory Group (eds) *Transformation of Our Food Systems: The Making of a Paradigm Shift*, Berlin: Zukenftsstiftung Landwirtschaft and the Biovision Foundation, pp 9–16.

Hirsch, J. (2022) 'Broad agriculture coalition files federal complaint against John Deere, demanding the right to repair their own tractors', The Counter, [online] 3 April, Available from: https://thecounter.org/john-deere-tractors-federal-complaint-right-to-repair-ftc/

Holmes, B. (2021) 'How has the pandemic strengthened the global food supply chain?' *Knowable Magazine*, 19 March, republished in The Counter, [online] 25 March, Available from: https://thecounter.org/pandemic-global-food-supply-chain-covid-19/

Holmes, D.E. (2017) *Big Data: A Very Short Introduction*, Oxford: Oxford University Press.

Holtslander, C. (2015) *Losing our Grip, 2015 Update: How Corporate Land Buy-up, Rising Farm Debt and Agribusiness Financing of Inputs Threaten Family Farms*, Washington, DC: National Farmers Union.

Howard, P. (2015) 'Intellectual property and consolidation in the seed industry', *Crop Science*, 55(6): 2489–95.

Howard, P. (2016) *Concentration and Power in the Food System: Who Controls What We Eat?* London: Bloomsbury Press.

Howard, P. (2018) 'Global seed industry changes since 2013', Phil Howard, [online] 31 December, Available from: https://philhoward.net/2018/12/31/global-seed-industry-changes-since-2018/

Howard, P. (2021) *Concentration and Power in the Food System: Who Controls What We Eat?* Revised edition, London: Bloomsbury Press.

Howard, P. (2022) 'Op-ed: Fake meats won't solve the climate crisis', Civil Eats, [online] 7 April, Available from: https://civileats.com/2022/04/07/op-ed-fake-meat-wont-solve-the-climate-crisis/

Howard, P., Ajena, F., Yanaoka, M., and Clarke, A. (2021) '"Protein" industry convergence and its implications for resilient and equitable food systems', *Frontiers in Sustainable Food Systems*, [online] 16 August, DOI: 10.3389/sfufs.2021.684181.

Humbird, D. (2021) 'Scale-up economics for cultured meat', *Biotechnology and Bioengineering*, 118(8): 3239–50.

IAASTD: International Assessment of Agricultural Knowledge, Science and Technology for Development (2009) *Agriculture at a Crossroads – Global Report*, Nairobi: United Nations Environment Programme.

IATP: Institute of Agriculture and Trade Policy (2020) *Revisiting Crisis by Design: Three Decades of Failed Farm Policy*, Minneapolis, MN: IATP.

ILO: International Labour Organisation (2021) *World Employment and Social Outlook: The Role of Digital Labour Platforms in Transforming the World of Work*, Geneva: ILO.

IPCC: Intergovernmental Panel on Climate Change (2018) *Special Report. Global Warming of 1.5C*. DOI: 10179/781009157940k.

IPCC (2020) *Special Report: Climate Change and Land*, IPCC [online], Available from: https://www.ipcc.ch/site/assets/uploads/sites/4/2020/02/SPM_Updated-Jan20.pdf

IPCC (2021) *AR6 Climate Change 2021: The Physical Science Basis*, IPCC [online], Available from: https//www.ipcc.ch/report/ar6/wg1/

IPCC (2022) *Sixth Assessment Report*, IPCC [online] Available from: https://www.ipcc.ch/assessment-report/ar6.

IPES-Food: International Panel of Experts on Sustainable Food Systems (2017) *Too Big to Feed: Exploring the Impacts of Mega-mergers, Concentration and Consolidation of Power in Agri-food Systems*, IPES-Food [online], Available from: http://www.ipes-food.org/publications

IPES-Food (2020) 'Communique: Covid-19 and the crisis in food systems: symptoms, causes and potential solutions', IPES-Food [online], Available from: https://www.ipes-food.org/pages/covid19

IPES-Food (2021a) *Briefing Note 1. On The Governance of Food Systems. An 'IPCC for Food?' How the UN Food Systems Summit is Being Used to Advance a Problematic New Science-Policy Agenda*, Lead authors: J. Clapp, M. Anderson, M. Rahmanian, and S. Montsalve Suarez, *IPES-Food* [online], Available from: https://www.ipes-food.org/_img/upload/files/GovBrief.pdf

IPES-Food (2021b) 'Withdrawal from the UNFSS: Memo from the IPEs-Food panel', 26 July, IPES-Food [online], Available from: https://www.ipes-food.org/_img/upload/files/UNFSS%20Withdrawal%20Statement.pdf

IPES-Food (2022a) 'The politics of protein, Examining the claims about livestock, fish, "alternative proteins" and sustainability', IPES-Food [online], Available from: https://www.ipes-food.org/publications

IPES-Food (2022b) 'Another perfect storm', https://ipes-food.org/pages/foodpricecrisis

IPES-Food and ETC Group (2021) *A Long Food Movement: Transforming Food Systems by 2045*, IPES-Food [online], Available from: http://www.ipes-food.org/pages/LongFoodMovement

Janssen, S.J.C., Porter, C.H., Moore, A.D., Athanasiadis, I.N., Foster, I., Jones, J.W., and Antle, J.M. (2017) 'Towards a new generation of agricultural system data, models and knowledge products', *Agricultural Systems*, 155: 200–12.

Jasanoff, S. and Hurlbut, J.B. (2018) 'A global observatory for gene editing', *Nature*, 555: 435–7.

Jeffrey, C. and Dyson, J. (2021) 'Geographies of the future: prefigurative politics', *Progress in Human Geography*, 45(4): 641–58.

Jeffries, A. (2020) 'During the pandemic, Grubhub should be thriving. It's not', The Markup, [online] 27 May, Available from: https://themarkup.org

Jia, H. (2020) 'Merger of Sinochem and ChemChina, long rumoured, is confirmed', *Chemical and Engineering News*, [online] 98(35), Available from: https://cen.acs.org/business/mergers-&-acquisitions/Merger-Sinochem-ChemChina-long-rumored/98/i35?ref=search_results

Jochems, C., van der Valk, J., Staffau, F.R., and Baumans, V. (2002) 'The use of foetal bovine serum: ethical or scientific problem?' *Alternatives to Laboratory Animals*, 30(2): 219–27.

Jolly, J. (2021) 'Deliveroo to withdraw from Spain in face of "rider law"', The Guardian, [online] 31 July. Available at: https://www.theguardian.com/business/2021/jul/30/deliveroo-unveils-plans-to-pull-out-of-spain-in-wake-of-rider-law#:~:text=Deliveroo%20has%20announced%20plans%20to,economy%20workers%20greater%20employment%20rights

Jolly, J. (2022) 'Board of JustEatTakeaway should be fired says key shareholder', The Guardian, [online] 26 April, Available at: https://www.theguardian.com/business/2022/apr/25/just-eat-takeaway-investors-cat-rock-capital-share-price

Juma, C. (1989) *The Gene Hunters*, London: Zed Books.

Karmaus, A.L. and Jones, W. (2020) 'Future foods symposium on alternative proteins: workshop proceedings', *Trends in Food Science and Nutrition*, DOI: 10.1016/jtifs.2020.06.018.

Kay, I.E. (1993) *The Molecular Vision of Life: Caltech, The Rockefeller Foundation, and the Rise of The New Biology*, New York: Cambridge University Press.

Kazan, O. (2019) 'The coming obsolescence of animal meat', *The Atlantic*, 16 April.

Keller, E.F. (1995) *Refiguring Life: Metaphors of Twentieth Century Biology*, New York: Columbia University Press.

Kelloway, C. (2020a) 'Can Amazon's new grocery store challenge mainstream supermarkets?', Food and Power, [online] 10 September, Available from: https://www.foodandpower.net/latest/2020/09/10/can-amazons-new-grocery-store-challenge-mainstream-supermarkets

Kelloway, C. (2020b) 'Restaurants are at the mercy of delivery apps but can they survive without them?' Civil Eats, [online] 20 June, Available from: http://civileats.com/20/06/23/restaurants-are-at-the-mercy-of-delivery-apps-but-can-they-survive-without-them/

Kelloway, C. (2021a) 'Tech giants burn cash on grocery delivery, but at what expense?', Food and Power, [online] 13 May, Available from: https://www.foodandpower.net/latest/uber-gopuff-vc-grocery-delivery-5-2021

Kelloway, C. (2021b) 'What Biden's Executive Order on Competition means for food systems', Food and Power, The Open Markets Institute, [online] 15 July, Available from: https://www.foodandpower.net/latest/biden-competition-eo-july-2021

Kelloway, C. and Miller, S. (2019) 'Food and power: addressing monopolization in America's food system', Open Markets Institute, [online] 27 March, Available from: https://www.openmarketsinstitute.org/publications/food-power-addressing-monopolization-americas-food-system

Kelso, D.D.T. (2003) 'Recreating democracy', in R.A. Schurman and D.D.T. Kelso (eds) *Engineering Trouble: Biotechnology and its Discontents*, Berkeley, CA: University of California Press, pp 239–53.

Kenney, M. (1986) *Biotechnology: The University-industrial Complex*, New Haven, CT: Yale University Press.

Kenney, M. (1998) 'Biotechnology and the creation of a new economic space', in A. Thachray (ed) *Private Science: Biotechnology and the Rise of the Molecular Sciences*, Philadelphia, PN: University of Pennsylvania Press, pp 131–43.

Kernecker, M., Knierim, A., Wurbs, A., Kraus, T., and Borges, F. (2019) 'Experience versus expectation: farmers' perceptions of smart farming technologies for cropping systems across Europe', *Precision Agriculture*, DOI: s11119-019-09651-z.

Kevles, D.J. (1998) '*Diamond v. Chakrabarty* and beyond: the political economy of the patenting of life', in A. Thackray (ed) *Private Science: Biotechnology and the Rise of The Molecular Sciences*, Philadelphia: University of Pennsylvania Press, pp 65–79.

Key, N. and Roberts, M.J. (2007) *Economic Research Report Number 51: Commodity Payments, Farm Business Survival, and Farm Size Growth*, Washington DC: USDA/ERS.

Kitchin, R. and Dodge, M. (2011) *Code/space: Software and Everyday Life*, Cambridge, Mass: MIT Press.

Kitchin, R. and McArdle, G. (2016) 'What makes big data, big data? Exploring the ontological characteristics of 26 datasets', *Big Data and Society* 3(1), DOI:10.1177/2053951716631130.

Klerkx, L. and Rose, D.C. (2020) 'Dealing with the game-changing technologies of Agriculture 4.0: how do we manage diversity and responsibility in food system transition?' *Global Food Security*, 24, DOI:10.1016/j.gfs.2019.100347.

Klerkx, L., Jakku, E., and Labarthe, P. (2019) 'A review of social science on digital agriculture, smart farming and agriculture 4.0', *NJAS – Wageningen Journal of Life Sciences*, 90–91 1–16, DOI:10.1016/njas.2019.100315.

Kloppenburg Jr., J.R. (1984) 'The social impact of biogenetic technology in agriculture: past and future', in G.M. Beradi and C.C. Geisler (eds) *The Social Consequences and Challenges of New Agricultural Technologies*, Boulder, CO and London: Westview Press.

Kloppenburg Jr., J.R. (1988) *First the Seed: The Political Economy of Plant Biotechnology* (1st edn), Cambridge: Cambridge University Press.

Kloppenburg Jr., J.R. (2004) *First The Seed: The Political Economy of Plant Biotechnology* (2nd edn), Madison, WI: University of Wisconsin Press.

Kneafsey, M., Maye, D., Holloway, L. and Goodman, M.K. (2021) *Geographies of Food: An Introduction*, London: Bloomsbury Academic.

Koch, L. (2019) 'Amazon, Walmart and Kroger lead US grocery sales online – how do they do it?', E-Marketer, [online] 17 June, Available from: https://www.emarketer.com/content/amazon-walmart-and-kroger-lead'us-grocery-sales-online-how-do-they-do-it

Kofler, N. and Kouzma, J. (2020) 'Before genetically modified mosquitoes are released, we need a better EPA', Boston Globe, 22 June.

Kuzma, J. (2019a) *Biotechnology Oversight Gets an Early Makeover by Trump's White House and The USDA: Part 1 – The Executive Order*, Raleigh, NC: Genetic Engineering and Society Center, North Carolina University.

Kuzma, J. (2019b) *Biotechnology Oversight Gets an Early Makeover by Trump's White House. Part 2 – The USDA/APHIS Rule*, Raleigh, NC: Genetic Engineering and Society Center, North Carolina University.

Kuzma, J. (2020) *Genetically Modified Mosquitoes Could be Released in Florida and Texas Beginning This Summer – Silver Bullet or Jumping The Gun?* Raleigh, NC: Genetic Engineering and Society Center, North Carolina University.

Kuzma, J. and Grieger, K. (2020) 'Community governance for gene edited crops', *Science*, 370(6519): 916–18.

Lachapelle, T. (2021) 'Grab's SPAC-to-riches plan creates $40 billion super-app', Bloomberg, [online] 14 April, Available from: https://www.bloomberg.com/opinion/articles/2021-04-13/grab-s-40-billion-spac-to-riches-plan-centers-on-super-apps

Lakhani, N. (2021) 'America's year of hunger: how children and people of colour suffered most', The Guardian, [online] 14 April, Available at: https://www.theguardian.com/environment/2021/apr/14/americas-year-of-hunger-how-children-and-people-of-color-suffered-most

Lamb, C. (2019) The Spoon Newsletter, [online] 13 June, Available from: https://thespoon.tech

Lamine, C. (2020) *Sustainable Agri-Food Systems: Case Studies in Transitions Towards Sustainability from France and Brazil*, London: Bloomsbury Academic.

Latour, B. (1992) 'Where are the missing masses? The sociology of a few mundane artefacts', in W.J. Bijker and J. Law (eds) *Shaping Technology/Building Society*, Cambridge, MA: MIT Press, pp 225–58.

Latour, B. (1993) *We Have Never Been Modern*, Brighton: Harvester Wheatsheaf.

Latour, B. (1994) 'On technical mediation: philosophy, sociology, genealogy', *Common Knowledge*, 3: 29–64.

Leclerc, R. (2019) 'We're only in the second innings of plant-based meat', Agfunder News, [online] 9 May, Available from: https://agfundernews.com/were-only-in-the-second-innings-of-plant-based-meats

Leclerc, R. and Tilney, M. (2014) 'AgTech is the new queen of the green', AgFunder News, [online] 4 January, Available from: https://agfundernews.com/agtech-new-queen-green

Lefrancois, T. (2021) 'Disease transmission from animals to humans has tripled over the last century', *Review*, Paris: Institute Polytechnique de Paris.

LeMieux, J. (2020) 'Florida approves mosquito release to curb viruses', Genetic Engineering and Biotechnology News, [online] 21 August, Available from: https://www.genengnews.com/news/florida-approves-mosquito-release-to-curb-spread-of-viruses/

Leszczynski, A. (2019) 'Spatialities', in J. Ash, R. Kitcin and A. Leszczynski (eds) *Digital Geographies*, London: Sage, pp 13–23.

Levidow, L. and Young, R.M. (eds) (1985) *Science, Technology and the Labour Process*, Vol. 2, London: Free Association Books.

Levidow, L., Carr, S., and Weild, D. (2000) 'Genetically-modified crops in the European Union: regulatory conflicts as regulatory opportunities', *Journal of Risk Research*, 3(3): 189–208.

Levidow, L., Nieddu, M., Vivien, F.-D., and Befort, N. (2019) 'Transitions towards a European bioeconomy: Life sciences versus agroecology trajectories', in G. Allaire and B. Daviron (eds) *Ecology, Capitalism and the New Agricultural Economy. The Second Great Transformation*, London: Rouledge, pp 181–203.

Lewis, T. (2020) *Digital Food: From Platform to Paddock*, London: Bloomsbury Academic.

Litchfield, J.H. (1983) 'Single cell protein', *Science*, 219: 11 February.

Long, T.B., Blok, V., and Coninx, I. (2016) 'Barriers to the adoption and diffusion of technological innovations for climate smart agriculture in Europe: evidence from the Nethherlands, France, Switzerland and Italy', *Journal of Cleaner Production*, 12: 9–21.

Longevity Science Panel (2021) 'The Covid-19 pandemic', [online], Available from: https://longevitysciencepanel.co.uk/landg-assets/longevity-panel/lsp_covid_6_october_2021_publish.pdf

Lupton, D. and Feldman, Z. (eds) *Digital Food Cultures*, London: Routledge.

Lynch, J. and Pierrehumbert, R. (2019) 'Climate impacts of cultured beef and beef cattle', *Frontiers in Sustainable Food Systems*, 3(5): DOI:10.3389/fsufs.2019.00005.

Lyons, K. (2021) 'Judge rules California Proposition 22 gig workers law is unconstitutional', The Verge, [online] 21 August, Available from: https://www.the verge.com/2021/8/21/22635286/judge-rules-california-prop-22-gig-workers-law-unconstitutional

MacDonald, J.M. (2019) 'Mergers in seeds and agricultural chemicals: what happened?', *Amber Waves*, Washington DC: USDA/ERS.

MacDonald, J.M. and Hoppe, R.A. (2017) 'Large family farms still dominate U.S. agricultural production', *Amber Waves*, Washington DC: USDA/ERS.

MacDonald, J.M., Hoppe, R.A., and Newton, D. (2018) *Three Decades of Consolidation in US Agriculture*, Washington DC: USDA/ERS [online], Available from: https://ers.usda.gov/webdocs/publications/88057/eib89.pdf?v=43172.

MacDonald, J.M., Law, J., and Mosheim, R. (2020) *Consolidation in U.S. Dairy Farming*, Washington DC: USDA/ERS.

Marsden, T.K., Flynn, A., and Harrison, M. (2000) *Consuming Interests: The Social Provision of Food*, London: UCL Press.

Marsden, T.K., Flynn, A., and S. Thankappen (2010) *The New Regulation and Governance of Food: Beyond the Food Crisis?* London: Routledge.

Marshall, J.M. and Akbari, O.S. (2018) 'Can CRISPR-based gene drives be confined to the wild? A question for molecular and population biology', *ACS Chemical Biology*, 13(2): 424–30.

Marston, J. (2020a) 'Fee caps can't save restaurants from third delivery practices', The Spoon, [online] 17 March, Available from: https://thespoon.tech/fee-caps-cant-save-resaurants-from-third -delivery-practices/

Marston, J. (2020b) 'Prop 22's success has unsettling implications for third-party delivery's power', The Spoon, [online] 5 November, Available from: https://the spoon.tech/prop-22's-success-has-unsettling-implications-for-third-party-delivery's-power

Marston, J. (2021a) 'The next big tech for the virtual food hall', The Spoon [online], Available from: https://thespoon.tech/the-next-big-tech-for-the-virtual-food-hall/

Marston, J. (2021b) 'Grocery, meet the ghost kitchen', The Spoon, [online] 14 July, Available from: https://thespoon.tech/grocery-meet-the-ghost-kitchen/

Marston, J. (2022) 'Data snapshot: alt-protein drove innovative foods US$ 4.5 billion funding in 2021', AgFunder News, [online] 27 April, Available from: https://agfundernews.com/innovative-food-nets-4-8b-in-2021-thanks-to-alt-protein-data-snapshot

Martyn-Hemphill, R. (2020) 'AgFunder Agri-Foodtech Investment Report – 2019', AgFunder News, [online] 25 February, Available from: https://agfundernews.com/research/agfunder-agrifood-tech-investing-report-2019/

Mascarenhas, M. and Busch, L. (2006) 'Seeds of change: intellectual property rights, genetically modified soybeans, and seed saving in the United States', *Sociologia Ruralis*, 46(2): 122–38.

Mason, S. (2018) 'High score, low pay: why the gig economy loves gamification', The Guardian, [online] 26 November, Available from: https://www.theguardian.com/business/2018/nov/20/high-score-low-pay-gamification-lyft-uber-drivers-ride-hailing-gig-economy

Mau, S. (2019) *The Metric Society: On the Quantification of The Social*, Cambridge: Polity Press.

McMichael, P. (2005) 'Corporate development and the corporate food regime', in F.H. Buttel and P. McMichael (eds) *New Directions in the Sociology of Global Development: Research in Rural Sociology and Development*, Vol. 11, Amsterdam: Elsevier, pp 265–99.

Medina, R.F. (2018) 'Gene drives and the management of agricultural pests', *Journal of Responsible Innovation*, 5(1): S255–62.

Mercatus/Incisiv (2020) 'E-grocery's new identity: the pandemic's lasting impact on US grocery shopping behaviour', Mercatus, [online], Available from: https://www.mercatus.com

Mercier, S. (2019) 'The current state of the U.S. dairy industry', Dairy Herd Management, [online] 6 November, Available from: https://www.dairyherd.com/news-news/business-markets/milk-prices/current-state-us-dairy-industry

Miles, C. (2019) 'The combine will tell the truth: on precision agriculture and algorithmic capitalism', *Big Data and Society*, 6(1): 1–12.

Montenegro de Wit, M. (2019) 'Gene driving the farm: who decides, who owns and who benefits?' *Agriculture and Sustainable Food Systems*, 43(9): 1054–74.

Montenegro de Wit, M. (2020a) 'Democratising CRISPR? Stories, practices and politics of science and governance on the agriculture gene editing frontier', *Elementa: Science for the Anthropocene*, 8(1), DOI: 10.1525/elementa.405.

Montenegro de Wit, M. (2020b) 'Who owns CRISPR?' *Elementa: Science for the Anthropocene*, 8(1), DOI:10.1525/elementa.405.

Montenegro de Wit, M. (2020c) 'How the new biotech rule will foster distrust with the public and impede the progress of science' *The Conversation*, 1 June.

Montenegro de Wit, M., Canfield, M., and Iles, A. (2021a) 'Weaponising science in global food policy', Santa Cruz: Inter Press Service News Agency, 25 June.

Montenegro de Wit, M., et al (2021b) 'UN Food System Summit plants corporate solutions and ploughs under people's knowledge', Agroecology Research Action Collective, [online] 16 July, Available from: https://agroecologyresearchaction.org/peoplesknowledge/

Mooney, P. (2015) 'The changing agribusiness climate: corporate concentration, agricultural inputs, innovation and climate change', *Canadian Food Studies*, 2(2): 117–25.

Moore, M. and Tambini, D. (eds) *Digital Dominance. The Power of Google, Amazon, Facebook and Apple*, New York, NY: Oxford University Press.

Moran, G. (2021) 'The food system's carbon footprint has been vastly underestimated', Civil Eats, [online] 30 June, Available from: https://civileats.com/2021/06/30/the-food-systems-carbon-footprint-has-been-vastly-underestimated/

Morgan, K., Marsden, T.K., and Murdoch, J. (2008) *Worlds of Food: Place, Power and Provenance in The Food Chain*, Oxford: Oxford University Press.

NASEM: National Academies of Sciences, Engineering and Medicine (2016) *Gene Drives on the Horizon: Advancing Science, Navigating Uncertainty, and Aligning Research With Public Values*, Washington, DC: National Academies Press.

Nasti, R.A. and Voytas, D.F. (2021) 'Attaining the promise of plant gene editing technology at scale', Colloquium paper, *Proceedings of the National Academy of Sciences*, 118(22), DOI: 10.1073/pnas.2004846117.

Nehring, R., Sauer, J., Gillespie, J., and Hallahan, C. (2016) 'United States and European dairy farms: where is the competitive edge?' *International Food and Agribusiness Management Review*, 19(Issue B): 210–31.

Newby, H. (1983) 'The sociology of agriculture: towards a new rural sociology', *Annual Review of Rural Sociology*, 9: 67–81.

Newell, P. (2017) 'Contested landscapes: the global political economy of climate-smart agriculture', *Journal of Peasant Studies*, 45(1): 108–29.

NHGRI: National Human Genome Research Institute (2017) 'How does genome editing work?', [online], Available from: https://www.genome.gov/about-genomics/policy-issues/what-is-Genome-Editing

OTA: Office of Technology Assessment (1981) *Impacts of Applied Genetics*, Washington, DC: US Congress.

OTA: Office of Technology Assessment (1984) *Commercial Biotechnology: An Industrial Analysis*, Washington, DC: US Congress.

Otway, C. (2020) 'Out of sight, out of mind: inside London's "dark kitchens"', EastLondonlines, [online], Available from: https://www.eastlondonlines.co.uk/2020/04/behind-your-dinner-deliveroos-dark-kitchens/

Partridge, J. (2021a) 'Amid healthy orders can JustEat deliver on share price, too?' The Guardian, [online] 15 August, Available from: https://www.theguardian.com/business/2021/aug/15/despite-healthy-orders-can-just-eat-deliver-on-share-price-growth

Partridge, J. (2021b) 'Getir shows its appetite for growth with Weezy deal', The Guardian, [online] 24 November, Available from: https://www.theguardian.com/business/2021/nov/23/rapid-delivery-service-getir-to-buy-uk-rival-weezy

Peet, E.R. and Peet, R. (2020) 'COVID-19: disease of global capitalism, excursions into spatial epidemiology', *Human Geography*, 13(3): 318–21.

Pellica-Harris, A., de Gama, N., and Ravishankar, M.N. (2020) 'Postcapitalist precarious work and those in the drivers' seat: exploring the motivations and lived experiences of Uber drivers in Canada', *Organization*, 27(1): 36–59.

Perutz, M. (1985) 'The birth of protein engineering', *New Scientist*, 1460, 13 June.

Peschard, K. and Randeria, S. (2020) '"Keeping seeds in our hands". The rise of seed activism', *Journal of Peasant Studies*, 47(4): 613–47.

Pixley, K.R., Falk-Zepeda, J.B., Giller, K.E., Glenna, L.L., Gould, F., Mallory-Smith, C.A., Stelly, D.M., and Stewart Jr., C.N. (2019) 'Gene editing, gene drives and synthetic biology: will they contribute to disease-resistant crops and who will benefit?' *Annual Review of Phytopathology*, 57: 165–88.

Plume, K. (2014a) 'High-tech U.S. farm machines harvest Big Data, reap privacy worries', Reuters, [online] 10 April, Available from: https://www.reuters.com/article/usa-farming-data-idINL2N0N11U720140409

Plume, K. (2014b) 'The big data bounty: US start-ups challenge agribusiness giants', Reuters, [online] 6 October, Available from: www.reuters.com/article/uk-usa-farming-startups-idUKKCN0HX0BK20141008

Poinski, M. (2019) 'Just Egg cracks the substitute category wide open', FoodDive, [online] 21 February, Available from: https://www.fooddive.com/news/just-egg-cracks-the-substitute-category-wide-open/548286/

Poinski, M. (2021a) 'Aleph Farms unveils the world's first cell-based ribeye steak', FoodDive, [online] 10 February, Available from: https://www.fooddive.com/news/aleph-farms-unveils-worlds-first-cell-based-ribeye-steak/594830/#:~:text=Israeli%20cell%2Dbased%20meat%20producer,cell%20cultivation%20and%203D%20bioprinting

Poinski, M. (2021b) 'Cell-based Meat-Tech 3D starts paperwork for US IPO', FoodDive, [online] 8 March, Available from: https://www.fooddive.com/news/cell-based-meat-tech-3d-starts-paperwork-for-us-ipo/587864/

Poinski, M. (2021c) 'The Better Foods Co. unveils mycoprotein fermentation line', FoodDive, [online] 8 June, Available from: https://www.fooddive.com/news/the-better-meat-co-unveils-mycoprotein-fermentation-line/601332/

Poinski, M. (2021d) 'Upside Foods develops animal-free growth medium for cell-based meat', FoodDive, [online] 9 December, Available from: https://www.fooddive.com/news/upside-foods-develops-animal-free-growth-medium-for-cell-based-meat/611218/

Poinski, M. (2022a) 'As plant-based meat growth stalls, what does this mean for the category?' FoodDive, [online] 28 February, Available from: https://www.fooddive.com/news/as-plant-based-meat-growth-stalls-what-does-it-mean-for-the-category/619437/

Poinski, M. (2022b) 'Beyond Meat shares drop as as jerky costs drive $100M loss', FoodDive, [online] 12 May, Available from: https://www.fooddive.com/news/beyond-meat-earnings-jerky-100m-loss/623635/

Poinski, M. (2022c) 'Beyond Meat accused of unfair competition because of misleading claims in a new lawsuit', FoodDive, [online] 3 June, Available from: https://www.fooddive.com/news/beyond-meat-unfair-competition-lawsuit-don-lee-farms/624824/

Poinski, M. (2022d) 'Mars partners with Perfect Day on animal-free chocolate', FoodDive, [online] 16 June, Available from: https://www.fooddive.com/news/mars-and-perfect-day-partner-on-animal-free-chocolate/#:~:text=Mars'%20CO2COA%20chocolate%20sources%20Rainforest,wrapped%20in%20paper%2Dbased%20packaging

Poinski, M. (2022e) 'Beyond Meat lays off 200 employees and slashes its revenue outlook', FoodDive, [online] 14 October, Available from: https://www.fooddive.com/news/beyond-meat-layoffs-cut-revenue-outlook/634141/

Pope, M. and Sonka, S. (2020) 'Quantifying the benefits of on-farm digital technologies', farmdoc daily, [online] 4 March, Available from: https://farmdoc.illinois.edu/2020/03/quantifying-the-economic-benefits-of-on-farm-digital-technologies.html

Poppe, K. (2016) *Big Opportunities for Big Data in Food and Agriculture*, OECD Workshop, February, Wageningen, Netherlands: Wageningen University and Research.

Power, M.E. (2021) 'Synthetic threads through the web of life', *Proceedings of the National Academy of Sciences*, 118(22), DOI:10.1073/pnas.2004833118.

Power, M., Doherty, B., Pybus, K., and Pickett, K. (2020) 'How Covid has exposed inequalities in the UK food system: the case of UK food and poverty', *Emerald Open Research*, 2(11), DOI: 10.35241/emeraldopenres.13539.

Rankin, J. (2021) 'EU employment rights crackdown to end gig economy "free-for-all"', The Guardian, [online] 10 December, Available from: https://www.theguardian.com/business/2021/dec/09/gig-economy-workers-to-get-employee-rights-under-eu-proposals

Redman, R. (2020) 'Online grocery to more than double its market share', Supermarket News, [online] 8 September, Available from: https://www.supermarketnews.com/online-retail/online-grocery-more-double-market-share-2025

Reed, M. (2002) 'Rebels from the Crown down: the organic movement's revolt against agricultural biotechnology', *Science as Culture*, 11(4): 481–504.

Reed, M. (2008) 'The rural arena: the diversity of protest in rural England', *Journal of Rural Studies*, 24(2): 209–18.

Reynolds, M. (2018) 'The clean meat industry is racing to ditch its reliance on foetal blood', WIRED UK, [online] 20 March, Available from: https://www.wired.co.uk/article/scaling-clean-meat-serum-just-finless-foods-mosa-meat?utm_source=onsite-share&utm_medium=email&utm_campaign=onsite-share&utm_brand=wired-uk

Ripple, W.J., Wolf, C., Newsome, T.M., Barnard, P., and Moomar, W.R. (2020) 'World scientists warning of climate emergency', *BioScience*, 70: 8–12.

Roberts, D. (2020) 'DoorDash IPO is "most ridiculous of 2020" and "holds no value": analyst', Yahoo! Finance, [online] 10 December, Available from: https://finance.uk.yahoo.com/news/door-dash-ipo-is-most-ridiculous-of-2020-and-holds-no-value-analyst-125054305.html?guccounter=1&guce_referrer=aHR0cHM6Ly93d3cuZ29vZ2xlLmNvbS8&guce_referrer_sig=AQAAADMGYi0-V4LkCxuELGdZv2P5Xk7B-LuiTE8GTf2QWtTnvrNYPuDNIubaN9X837hOVOgQPg0bsOUg7kFYwArDc0zhqM0k8AOACnRdRf6kl7JXEDAom6smTjL1NhfPOPNfeqw3SEk221LGNJgrWAiVYu64GWEjly-Q9tWKK0c_MkAc

Rodrigues, V. (2008) 'Governance of material transfer agreements', *Technology in Society*, 30(2): 222–8.

Rogers, B. (2018) 'How Agtech has quietly transformed agriculture in the downturn', AgFunder News, [online] 27 August, Available from: https://agfundernews.com/agtech-quietly-transformed-agriculture-downturn

Rose, D.C. and Chivers, J. (2018) 'Agriculture 4.0: broadening responsible innovation to an era of smart farming', *Frontiers in Sustainable Food Systems*, 21 December, DOI:10.3389/fsufs.2018.00087.

Rose, D.C., Morris, C., Lobley, M., Winter, M., Sutherland, W.J., and Dicks, L.V. (2018) 'Exploring the spatiality of technological and user re-scripting: the case of decision support tools in UK agriculture', *Geoforum*, 89: 11–18.

Rotz, S., Duncan, E., Small, M., Botschner, J., Dara, R., Mosby, I., and Reed, M. (2019a) 'The politics of digital agriculture technologies: a preliminary review', *Sociologia Ruralis*, 59(2): 203–12.

Rotz, S., Gravely, E., Mosby, I., Duncan, E., Finnis, E., Morgan, M., Neufeld, E.T., Nisson, A., Pant, L., Shalla, V., and Frazer, E. (2019b) 'Automated pastures and the digital divide: how agricultural technologies are shaping labour and rural communities', *Journal of Rural Studies*, 68: 112–22.

Ruehl, M. and Palma, S. (2021) 'A driven entrepreneur grabs his opportunity', Financial Times, 9 April.

Rushe, D. (2021) 'Big Tech: the threat from a handful of firms that made billions during the pandemic', The Guardian, 31 July.

Santo, R. and Clinton, S. (2017) *Redefining Protein: Adjusting Diets to Protect Public Health and Conserve Resources*, Health Care Without Harm, [online], Available from: https://noharm-uscanada.org/RedefiningProtein

Santo, R., Kim, B., Goldman, S., Dutkiewicz, J., Biehl, E., Bloem, M., Neff, R., and Nachman, K. (2020) 'Considering plant-based meat substitutes and cell-based meats; a public health and food systems perspective', Review paper, *Frontiers in Sustainable Food Systems*, 31 August, DOI: 10.3389/fsufs.2020.00134.

Schimmelpfennig, D. (2016) *Farm Profits and Adoption of Precision Agriculture – Economic Research Report No. 217*, Washington, DC: USDA/ERS.

Schimmelpfennig, D. and King, J. (2006) 'Mergers, acquisitions and flows of intellectual property', in R.E. Evenson and V. Santaniello (eds) *International Trade and Policies for Genetically Modified Products*, Wallingford, UK: CABI Publishing, pp 97–109.

Schmaltz, R. (2017) 'What is Precision Agriculture?' Agfunder News, [online] 24 April, Available from: www.https://agfundernews.com/what-is-precision-agriculture

Schurman, R. and Munro, W.A. (2003) 'Making biotech history: social resistance to agricultural biotechnology and the future of the biotechnology industry', in R. Schurman and D.D.T. Kelso (eds) *Engineering Trouble: Biotechnology and its Discontents*, Berkeley, CA: University of California Press, pp 111–29.

Scoones, I. (2022) 'Livestock, methane and climate change: The politics of global assessments', *WIREs Climate Change*, [online] 27 May, DOI:10.1002/wcc.790.

Scrinis, G. (2008) 'On the origins of nutritionism', *Gastronomica*, 8(1): 39–48.

Sexton, A. (2014) *Accumulation by Simulation: Cultured Meat, Food Security and the Materialisation of Biopolitics: Upgrade Report*, London: King's College London and the Economic and Social Research Council.

Sexton (2018) 'Bringing cultured meat to market: technical, socio-political and regulatory challenges in cellular agriculture', *Trends in Food Science and Technology*, 78: 155–66.

Sexton, A., Garnett, T., and Lorimer, J. (2019) 'Framing the future of food: the contested promises of alternative proteins', *Environment and Planning E: Nature and Space*, 2(1): 47–72.

Sharma, S. (2020) *Milking the Planet. How Big Dairy is Heating Up The Planet and Hollowing Out Our Rural Communities*, Minneapolis, MN: IATP: Institute of Agriculture and Trade Policy.

Sherkow, J.S. (2022) 'Immaculate conception? Priority and invention in the CRISPR patent dispute', *CRISPR Journal*, 5(2), DOI:10.1089/crispr.2022.0033.

Shukla-Jones, A., Friedrichs, S., and Winickoff, D.E. (2018) 'Gene editing in international context: science, economic and social issues across sectors', OECD Science, Technology and Industry Working Papers 2018/04, Paris: OECD.

Skotstad, E. (2020) 'United States relaxes rules for biotech crops', *Science*, DOI:10.1126/science.abc8305.

Sonka, S. (2016) 'Big data characteristics', *International Food and Agri-business Management Review*, 19(Issue A): 7–11.

Sonka, S. and Cheng, Y.-T. (2015) 'A Big Data Revolution: what would drive it?', farmdoc daily, [online] 5: 223, Available from: https://farmdoc.illinois.edu/2015/12/a-big-data-revolution-what-would-drive-it.html

Spackman, C. (2019) 'The problem with lab-grown meat', Future Tense, [online] 7 May, Available from: https://slate.com/technology/2019/05/lab-grown-meat-food-agriculture-system.html

Srnicek, N. (2017a) *Platform Capitalism*, Cambridge: Polity Press.

Srnicek, N. (2017b) 'Smash the tech monopolies now. Soon it will be too late', The Guardian, 30 August.

Stallman, R. (2018) 'Why open source misses the point of free software', GNU, [online], Available from: https://www.gnu.org/phliosophy/open-source-misses-the-point.hmtl.en

Stanford Graduate School of Business (2017) 'Technology in agribusiness. Opportunities to drive value', White Paper, [online] August, Available from: https://www.gsb.stanford.edu/faculty-researcg/working-papers.

Steinbrecher, R.A. (2015) 'Genetic engineering in plants and the "New Breeding Techniques (NBTs)". Inherent risks and the need to regulate', EcoNexus Briefing, [online], Available from: www.econexus.info/publication/genetic-engineering-plants-and-new-breeding-techniques

Stephens, N., Sexton, A., and Driessen, C. (2019) 'Making sense of making meat: key moments in the first 20 years of tissue-culture engineering muscle to make food', *Frontiers in Sustainable Food Systems*, 10 July, DOI:10.3389/fsufs.2019.00045.

Stephens, N., Di Silvio, L., Dunsford, I., Ellis, M., Glencross, A., and Sexton, A. (2018) 'Review. Bringing cultural meat to market: technical, socio-political and regulatory challenges for cellular agriculture', *Trends in Food Science*, 78: 155–66.

Stine, L. (2019) 'Meet the farmer: Lynn Rohrscheib is staying true to her farmer roots', AgFunder News, [online] 19 November, Available from: https://agfundernews.com/meet-the-farmer-lynn-rohrscheib-is-staying-true-to-her-farming-roots

Stoller, M. (2020) 'Counterfeit capitalism, food delivery apps and the attack on franchising', BIG, Matt Stoller, [online] 23 July, Available from: https://mattstoller.substack.com/p/counterfeit-capitalism-food-delivery

Strauss, D., Bradshaw, T., and Mooney, A. (2021) 'Legal threats make for rough ride at Deliveroo', Financial Times (Companies and Markets), 26 March, p 10.

Sugar, R. (2018) '2018 – the year vegan junk food went mainstream', VOX, [online] 25 December, Available from: https://www.vox.com/the-goods/2018/12/25/18127863/vegan-beyond-impossible-burger

Sweney, M. and Wood, Z. (2021) 'Could robots take over from Ocado delivery drivers?' The Guardian, 16 April.

Swinburn, B.A., Kraak, V.I., Allender, S., Atkins, V.J., Baker, P.I., Bogard, J.R., Brinsden, H., Calvillo, A., De Schutter, O., Devarajan, R. et al (2019) 'Lancet Commission report: The global syndemic of obesity, undernutrition, and climate change', The Lancet, 393(10173): 791–846.

Swinburn, B.A. (2020) 'The agriculture-health nexus: a decade of paradigm progress but patchy policy actions', in H.R. Herren, B. Haerlin, and IAASTD+10 Advisory Group (eds) *Transformation of Our Food Systems: The Making of a Paradigm Shift*, Berlin: Zukenftsstiftung Landwirtschaft and the Biovision Foundation, pp 130–6.

Sykuta, M.E. (2016) 'Big Data in agriculture: property rights, privacy and competition in agricultural data services', *International Food and Agri-Business Management Review*, 19(issue A): 57–74.

Taylor, F.W. (1911) *The Principles of Scientific Management*, New York: Harper.

Toya, M., Kurawadwala, H., Bryne, B., Dowby, R., and Weston, Z. (2022) *Plant-based Meat: Anticipating 2030 Production Requirements*, Washington, DC: Good Food Institute.

Tubb, C. and Seba, T. (2019) 'Rethinking food and agriculture, 2020–2030. The second domestication of plants and animals, the disruption of the cow, and collapse of industrial livestock farming', *Industrial Biotechnology* 17(2): 57–72.

Ulmer, K.J. (1983) 'Protein engineering', *Science*, 219, 11 February.

United Nations (2021) 'Climate and environment', *United Nations News*, 9 August.

Van der Ploeg, J.D., Franco, J.C and Borras, S.M. (2015) 'Land concentration and land grabbing in Europe a preliminary analysis', *Canadian Journal of Development Studies*, 36(2): 147–62.

Vanloqueren, G. and Baret, P.V. (2009) 'How agricultural research systems shape a technology regime that develops genetic engineering but locks out agroecological innovation', *Research Policy*, 38(6): 971–83.

Veen, A., Barratt, T., and Goods, C. (2020) 'Platform capitalism's "app-etite" for control: a labour process analysis of food delivery work in Australia', *Work, Employment and Society*, 34(3): 386–406.

Wallace, R. (2020) 'Update: agriculture, capital and infectious diseases', in H.R. Herren, B. Haerlin, and IAASTAD+10 Advisory Group, *Transforming our Food Systems: The Making of a Paradigm Change*, Berlin: Zukunftsstififtung Landwirtschaft and the Biovision Foundation, pp 79–83.

Wallace, R. (2021) 'Planet earth', *New Internationalist*, January.

Waltz, E. (2010) 'Glyphosate resistance threatens Roundup Ready's hegemony', *Nature Biotechnology*, 28(6): 337–8.

Waltz, E. (2021) 'First genetically modified mosquitoes released in Florida', *Nature*, 593: 175–6, 3 May.

Watson, E. (2022) 'Beyond Meat CEO faces gruelling from analysts after posting grim Q1 figures', Food Navigator, [online] 12 May, Available from: https://www.foodnavigator.com/Article/2022/05/12/Beyond-Meat-CEO-faces-grilling-from-analysts-after-posting-grim-Q1-figures

Watts, M.J. (2004) 'Are hogs like chickens? Enclosure and mechanisation in two "white meat" filieres', in A. Hughes and S. Reimer (eds) *Geographies of Commodity Chains*, London and New York: Routledge, pp 39–62.

Weis, T. (2015) 'Meatification and the madness of the doubling narrative', *Canadian Food Studies*, 2(2): 296–303.

Weldon, T. (2021) 'Why Instacart might deliver an unusual IPO', The Motley Fool, [online] 13 April, Available from: https://fool.com/2021/04/13/why-instacart-might-deliver-an-unusual-ipo/

Weston, P. and Watts, J. (2021) 'The cow in the room: why is no one talking about farming at COP26?' The Guardian, [online] 9 November, Available from: https://www.theguardian.com/environment/2021/nov/09/the-cow-in-the-room-why-is-no-one-talking-about-farming-at-cop26-aoe?CMP=share_btn_link

White, T. (2021) 'First GMO mosquitoes to be released in the Florida Keys', Raleigh, NC: Genetic Engineering and Society Center, North Carolina State University.

Wiens, K. (2015) 'New high-tech farm equipment is a nightmare for farmers', WIRED, [online], Available from: https://www.wired.com/2015/02/new-high-tech-farm-equipment-nightmare-farmers/

Wilkinson, J. (2019) *Grande Fuerzas, Global Tendencies, and Rural Actors With Respect to SDG Goals*, Rome: FAO.

Winter, A., Finger, R., Huber, R., and Buchmann, N. (2017) 'Opinion: Smart farming is the key to developing sustainable agriculture', *Proceedings of the National Academy of Sciences*, 114(24): 1648–50.

Wiseman, L., Sanderson, J., Zhang, A., and Jakku, E. (2019) 'Farmers and their data: an examination of farmers' reluctance to share their data through the lens of the laws impacting smart farming', *NJAS: Wageningen Journal of Life Sciences*, 90–91, DOI: 10.1016/j.njas.2019.04.007.

Wolf, S. and Buttel, F.H. (1996) 'The political economy of precision agriculture', *American Journal of Agricultural Economics*, 78(5): 1269–94.

Wolfort, S., Ge, L., Verdouw, C., and Bogaadt, M.J. (2017) 'Big data in smart farming: a review', *Agricultural Systems*, 153: 69–80.

Wolt, J.D. and Wolf, C. (2018) 'Governance perspectives for regulation of genome edited crops in the US', *Frontiers in Crop Science*, 9: 2–12.

Wood, E. (2020) 'Alt-milky way. Starry investors pile in as Swedish brand rides wave to $20 bn. valuation', The Guardian, 20 July.

WEF: World Economic Forum (2018) 'Innovation with a purpose: the role of technology innovation in accelerating food system transformation', WEF, [online], Available from: https://www3.weforum.org/docs/WEF_Innovation_with_a_Purpose_VF-reduced.pdf

Wright, S. 1994) *Molecular Politics: Developing American and British Regulatory Policy for Genetic Engineering, 1972–1982*, Chicago: University of Chicago Press.

Yglesias, M. (2017) 'The real reason Amazon buying Whole Foods terrifies te opposition', VOX, [online] 20 June, Available from: https://www.reddit.com/policy-and-politics/2017/6/20/15824718/amazon-whole-foods-profit-margin

Young, R.M. (1979) 'Science is social relations', *Radical Science Journal*, 5: 65–129.

Yoxen, E. (1981) 'Life as a productive force: capitalising the science and technology of molecular biology', in L. Levidow and R.M. Young (eds) *Science, Technology and the Labour Process: Marxist Studies*, Vol. 1, London: CSE Books.

Zimmer, C. (2017). '"Gene drives" are too risky for field trials, scientists say', New York Times, 16 November.

Zwik, A. (2018) 'Welcome to the gig economy: neoliberal industrial relations and the case of Uber', *GeoJournal*, 83(4): 679–91.

Index

References to endnotes show both the page number and the note number (98n10)

3D printing 38, 39
7-Eleven 72
640 Labs 98n10

A

acquisitions *see* mergers and acquisitions (M&A)
Addison Lee 82
AGCO 16, 97n7
AgConnections 18
agri-biotechnology 9–10, 53–4, 89
  agro-ecosystem engineering 61–4
  analysis of 69–70
  and de-regulation 68–9
  failed promises 56–9
  first-generation plant biotechnologies 56
  gene driving *see* gene drives
  gene editing *see* gene editing
  genomics 59–61
  regulation and governance 64–6
  regulatory policies 67–8
  vertical integration and seed–chemical complex concentration 54–5
agricultural knowledge and innovation system (AKIS) 33
AgriEdge Excelsior 18
AgTech space 12
Ainsworth, C. 68
Akbari, O. 61
Albertsons 76
Aleph Farms 39, 41
Alexandrov, V. 79
Alibaba 15, 45, 79
alternative proteins 4, 9, 36–7, 90, 94
  and bio-mimicry 37–9
  corporate routes to mainstream markets for 39–41
  and feedstocks 49
  and GHG emissions 49
  industry, structuring of 42–3
  plant-based protein 37–8, 42, 44, 46, 48–9, 90, 94, 104n3
  and price parity 46–7
  promissory narratives 46–50
  and protein conglomerates 43–6
  and Substitutionism 4.0 50–2
Altimeter Capital Management 103n8
Amazon 72, 75, 79, 97n1, 103n4, 103n9
Amazon Fresh 73, 74
Amazon Marketplace 15
Amazon Web Services 13, 97n1
Amyris 43
analytic biology 100n12
anticipatory politics 92–3
app-based delivery 76–7
  companies 80
  food/meal delivery 77–9
  third-party delivery apps 76–7
  vertical integration in 77–8
Apple 15
Apple Pay 79
Appropriationism 5–6
Appropriationism 4.0 33–5
Archer-Daniels Midland (ADM) 43, 99n1
Arla 45
Asilomar conference (1975) 66
AstraZeneca 2
Australia 30
automation 29
  of datasets 13
  equipment 23
  impact on farm labourers 33
  in online food delivery 86
  in retailing 75
  significance of 99n15
Awesome Burger 44

B

Barwise, P. 15, 78, 80, 81
BASF 18, 57, 64
Bayer 2, 13, 17, 18, 57, 98n14

134

INDEX

Bayer-Monsanto 16, 17, 19–20, 34, 54 64, 102n8
Bear Flag 97n2
Ben & Jerry's 44
Benbrook, C.M. 57
Bene, C. 94
Benson Hill 43
Better Food Co. 99–100n4
Better Meat Company 44
Bevmo 74
Beyond Meat 37, 40, 42, 44, 46, 48, 100n7
Biden, J. 104n1
Big Data 6, 8, 32, 97n4, 99n17
 and alternative protein industry 42
 impact of 19
 and precision agriculture 13–14, 15
big-box grocery retailers 10, 73, 87, 90
bio-mimicry and technologies
 cellular agriculture 38–9
 plant-based protein 37–8
bio-printing 38, 39
biotechnologies
 first-generation 56
 significance of 51
 university–industrial complex 55, 65
 see also agri-biotechnology
Blackstone 100n16
Blue Apron 75–6
Blue River Technologies 16
Bogue, A. 5
Böhm, S. 72
Borden 45
Boyd, W. 51, 65, 69
Bradshaw, T. 37
Bronson, K. 19, 32, 33
Brown, E. 48
Brown, P. 39–40, 49
bundling 16, 17, 19, 78, 105n10
 see also vertical integration
Burdett, D. 18
Burger King 42
Buttel, F.H. 12, 22, 89
Butz, E. 22, 88, 98n1

C

capitalism, impact of 24–5, 32
Caraher, M. 84
Cargill 44, 46
Caribou Biosciences 55, 57
Carl's Jr. 42
Carolan, M. 1, 29, 31, 33, 34
Carson, R. 88
cell-growth serum 40–1
cellular agriculture 36, 38–9, 43, 100n8
cellular meat products 40, 47
charitisation 84
Charpentier, E. 55
Chef's Plate 75
ChemChina 18, 98n13

Chipotle 86
Chirps 99n1
Chobani 45
Cibus 60, 68
clean meat see cultured meat (clean meat)
Climate Corporation 8, 16, 17, 18, 97–8n10, 98n11
Climate FieldView data (Bayer-Monsanto) 16, 17, 20, 34
Cloud Kitchens 76, 77
CNH 97n7
Cochrane, W.W. 89
Cohen-Bayer patent (1980) 65, 103n23
Colin, N. 79
collagen-based scaffolds 40
concentrated animal feeding operations (CAFOs) 23
contract production, in livestock industry 23
Coordinated Framework for Regulation of Biotechnology 66
co-productions 6, 13, 27, 72
 see also precision agriculture
corporate power 20, 21, 28, 64–5, 69, 96
Corteva Agriscience 18, 55, 64
COVID-19 pandemic 10, 71, 72, 87, 96, 104n15, 104n16
 and diffusion of digital technologies and practices 86–7
 and food deliveries 81, 89–90
 impact on food retailers and delivery platforms 85
 impact on restaurant business 76
 and online retailers 73, 78
 as socio-natural collective 84–5
 as stress test 3
Cowen 80
Cox, D. 31
Coyle, D. 15, 19
CRISPR technology 55, 63
 advantages of 60
 CRISPR-Cas9 55, 59, 61, 67
 and gene editing 59–60
Cruise 86
Cubbage, S. 14–15
cultured meat (clean meat) 38, 39, 41, 44, 49–50, 51
 see also alternative proteins

D

Dairy Farmers of America 45–6
dairy sector 23–4
Danone 43, 44, 46
dark kitchens (ghost kitchens) 76–7, 78
dark restaurants 76–7
dark store model 72, 73, 74
Day, S. 12
Dean Foods 45
decision support tools 14, 27, 28, 29

Deliveroo 76, 77, 79–82, 103n10, 104n11 (Ch 6), 104n17 (Ch 6), 104n2 (Ch 7)
Deliveroo Editions 72, 77–8
Diamond vs. Chakrabarty (1980) 64, 65, 66, 101n1
digital and molecular technologies, convergence of 1–2, 17
digital farming *see* precision agriculture (PA)
digital markets and platforms 15–17
digitalisation of field crop production 8
  see also GPS-assisted technologies; precision agriculture (PA)
Dodge, M. 6, 35, 86
Domino's 86
Don Lee Farms 48
DoorDash Essentials 72, 76, 77, 78, 80, 81, 83
DoorMart 78
Doudna, J. 55
Dow 57
Dow-DuPont 19
Dunhill, P. 50
DuPont/Pioneer 57

E

EAT JUST 37–8, 41, 42–3, 51, 100n6
ecosystem engineering 59–60, 61–4, 89
Embracing Nature campaign 68
Enterra 99n1
ETC Group 14, 19, 63, 88, 97n5, 97n7
European Court of Justice (ECJ) 68
European Food Safety Authority (EFSA) 67
European Union (EU) 24
  adoption of precision agriculture in 26, 27
  alternative proteins industry in 42–3
  biotechnology in 55, 58, 59
  Common Agricultural Policy 24
  dairy farms in 24
  gig economy 82, 84
  grocery chains in 73
  online grocery delivery in 74, 80
  regulation of agri-biotechnology 67
  structural fault lines and change in 22
EVERY (formerly Clara Foods) 39
evolutionary economics 2, 7, 32, 88
  see also lock-in; path-dependence
Exo 99n1

F

Facebook 15
farm equipment industry 13–14, 16–17, 30–1, 34–5, 97n2
Farm Hack 31, 99n12
farm service platforms 15–17
farmers
  and data ownership 31
  'dis-embedding' of knowledge and authority of 34
  disempowerment of 30–1
  identity and autonomy of 29–30
  and knowledge production 29
  privacy of 30, 31
  subjectivities, and digital innovations 28–32
  and transgenic pest management systems 56–7
Farmobile 14, 31
farmOS 31
FarmShots, Inc. 18
Fassler, J. 38, 50, 51
Federal Communications Commission 99n9
Feehan, P. 93
fermentation-based alternative proteins 36–7, 39, 40, 43, 100n12
Fernandez-Cornejo, J. 57
foetal bovine serum (FBS) 40–1
Food Rocket 74, 78, 79
FoodDive 100n14
food/meal delivery 77–9
FreshDirect 75
Freshly 76
Fridge No More 74
Froggatt, A. 41
Furey, S. 84
Future Meat Technologies 41, 44

G

Gatik 86
gene drives 10, 54
  concerns 62–3
  and CRISPR 60
  and engineering agri-ecosystem 61–4
  and genomics and gene editing 59–61
  and pest control 64, 69–70, 88–9
  regulations 63, 69
  reversal or immunisation drives 63
  risks and dangers of 63–4, 69
  scope of 54
  and species extinction 62
  synthetic 62
  varieties 62
gene editing 10, 11, 54, 56, 89
  herbicide-resistant oilseed rape 60–1
  marketing of 61
  patent and licencing 55
  pleiotropy 102n16
  potentiality 59–60
  and precision agriculture 93
  regulations 68
  risks and harms 59
  techniques 60, 67–8
  see also CRISPR; gene drives
General Mills 45
Genetic Manipulation Advisory Group (GMAG) 103n24
genetically modified (GM)
  crops 57–8, 67, 102n10
  feedstocks 49
  mosquitoes 63, 69, 102n17

INDEX

regulations 66, 67, 103n26
seeds 17, 55, 57, 65
genetically modified organisms (GMOs)
   and cell-cultured heme 100n9
   commercialization of 67
   and gene editing, compared 67–8
   and genetic engineering 103n27
   labelling 66
genome sequencing 36, 54, 60
germplasm diversity 58
'Get big or get out' phrase IATP report 22
Getir 74, 103n3, 104n18
GHG emissions, food system-related 2–4
ghost kitchens *see* dark kitchens (ghost kitchens)
Gibson, J. 29, 54
gig economy 10, 72, 75, 76, 78, 80–2, 104n11, 104n14
Gingko BioWorks 43
Global Open Data for Agriculture and Nutrition initiative (GODAN) 31
global warming 94–5, 105n12
   and genome editing 61
   IPCC report on 3, 94
   mitigating 3–4, 41, 47, 61, 90, 94
   moral posturing of 20
Goldman Sachs 103n10
Good Catch 99n3
'good' farming practice, impact of digital innovations on 29–30
Good Food Institute (GFI) 38, 42, 48–9, 100n15
Google 15
Gopuff 74, 75, 78–9
Gorillas 74, 78, 87
GPS-assisted technologies 6, 12–13, 25, 77
Grab Holdings, Inc. 79
grocery retailers
   online grocery shopping 73–6
   use of automation and robotics 75
grocery retails
   kerb-side and store pickups 73–4
gross cash farm income (GCFI) 22
GrubHub 72, 76, 81, 103n6, 103n9
Guterres, A. 3, 92

H

Hampton Creek Foods *see* EAT JUST
Heinemann, J. 58
Heinrich Boll Foundation 63
Hellmanns 44
Hello Fresh 76
heme 39–40, 42, 46, 100n9
herbicide-resistant oilseed rape 60–1
Hibberd, Ex Parte 65
High Level Panel of Experts on Food Security and Nutrition (HLPE) 91, 92, 104n5, 105n8

Home Chef 76
home delivery system 73–6, 77, 78, 85–6
Howard, P. 18, 54, 88, 89
human-induced climate change 2–3
Humbird, D. 50

I

IAASTD 91–2
IBM 97n1
iFixit 30
Impossible Burger 39–40, 100n9
Impossible Foods 37, 39, 42, 46, 100n11, 101n19
Impossible Whopper 42
industrial agriculture 2, 4–5, 7
information and communication technologies (ICTs) 26–7, 36, 59
infrastructural inequalities in connectivity 26
inheritance, and gene drive 62
InnovaFeed 43, 99n1
innovation pathways 1–2
Instacart 75, 83, 103n3, 103n5
Institute for Agriculture and Trade Policy (IATP) 5, 6, 53, 88, 98n2
Intergovernmental Panel on Climate Change (IPCC) 2, 3, 90, 94, 104n3
International Convention for the Protection of New Plant Varieties (UPOV) 64
International Panel of Experts on Sustainable Food Systems (IPES-Food) 32, 49, 91, 93, 96, 104n5, 105n9, 105n10
'IPCC for Food' 93
Israel 39

J

Jaffe, G. 69
Jantz, D. 60
JBS Foods 4, 44–5, 46
Jiffy 74
Jochems, C. 41
John Deere 2, 13, 14, 16, 32, 89, 97n2, 97n66, 99n11
JOKR 74
Just Egg 38, 100n6
Just Mayo 38, 100n6

K

Kalibata, A. 92
Karma Kitchen 77
Kelloway, C. 74
Kelso, D.D.T. 66, 67
Kenney, M. 55
Kernecker, M. 26, 29
Kitchen United 76
Kitchin, R. 6
Kloppenburg Jr., J.R. 56
Knezevic, I. 32
knowledge production 29, 31, 34

Kroger 73, 76
Kubata 97n7
Kuzma, J. 69

L

Lactalis 45
land consolidation
   and bi-modal farm size structure 22–3
   and livestock industry 23–4
'last-mile' delivery 78, 86, 90
Latour, B. 27
Leclerc, R. 18
Leszczynski, A. 6
life cycle assessments (LCAs) 49
life science corporations 4, 7, 9–10, 14, 18, 55–6, 64, 68–9, 89
Lightlife Foods 45
livestock industry
   and climate emergency 2–3, 49, 58
   contract production in 23
   and land consolidation 23–4
   and precision agriculture (PA) 23–5
lock-in 7, 16, 18, 33
   concept of 2, 32, 88
   farmer 8
   technological 89
Lyft 83

M

MacDonald, J. 53–4, 88
Maple Leaf Foods 44, 45, 48
Marlow Foods 99n4
Mason, S. 83
Mau, S. 34
McDonald's 42
McPlant 42
meal kits 75–6
meat substitutes 4, 9, 37, 38–9
   see also alternative proteins
Meat-Tech 3D 39, 41
Memphis Meats *see* Upside Foods
Mercatus/Incisiv 74
mergers and acquisitions (M&A) 16, 17, 18, 43, 44, 46, 73, 89
microbial fermentation 37, 38
micro-fulfilment centres 87
Microsoft 15, 97n1
mid-sized cropland farms, disappearance of 22
Miles, C. 34
miniaturisation 6, 34–5, 36
Modern Meadow 39
molecular farming 36
molecular reductionism 50
Monde Nissin 99n4, 100n5
Monsanto 2, 8, 13, 16–18, 32, 57–8, 97n10, 98n11
Montenegro de Wit, M. 60
Mooney, P. 32
Morrisons 104n17
Mosa Meat 41
Motif FoodWorks 43, 46
multi-stakeholderism 92, 93
Myco Tech 43
mycoprotein 37

N

National Institutes of Health (NIH) 66
Nationally Determined Commitments (NDCs) 93, 94
Nestle 44, 45, 76
Netscape 98n12
new breeding technologies 55
North America 22, 26, 33, 99n15
Novameat 38
Nuno 86
Nutriati 43

O

Oatly 45
Ocado 75, 86
Ocean Hugger Foods 99n3
'on-demand' home grocery deliveries 72
Open Ag Data Alliance (OADA) 31
Open Philanthropy 50
OpenTable 76
OSI Group 100n11
Oxbotica 86
Oxitec 61, 69, 102n17, 102n21
Ozo burgers 44

P

path-dependence 4, 7, 33
   concept of 2, 9, 32, 88
   technological 7, 9–10, 12, 13, 18, 32, 53, 89, 91, 93–5
Perdue, S. 22, 68, 89, 98n1
Perdue Farms 44
Perfect Day 38, 39, 40, 43
Pixley, K.R. 60
Plant-Based Food Association 100n15
plant-based protein 37–8, 42, 44, 46, 48–9, 90, 94, 104n3
   see also alternative proteins
Planterra 44
Plated 76
platform capitalism 10, 76, 80–2, 83
pleiotropy 102n16
Poinski, M. 41
Postgates 81
Power, M.E. 62–3
power, restructuring and relocation of 29, 30, 69
precision agriculture (PA) 8–9, 12–13, 21, 23
   adoption and diffusion for 25–7
   analysis of 19–20
   and Appropriationism 4.0 33–5
   and Big Data 13–14, 15
   digital support platforms for 14

INDEX

digitalisation 13–15
disempowerment of farmers by 30–1
and farmer identity, autonomy, and control 28–32
and farmer subjectivities 28–32
and livestock industries 23–5
and path dependence 32–3
platforms and consolidation in agricultural life sciences 17–19
platforms and digital markets for 15–17
re-scripting for 27–8
Precision BioScience 60
Precision Planting 16, 17, 97n10
privacy issues 30, 31
Proposition 22, of California 82–4, 86, 104n13

Q

Quorn 37, 99n4, 100n5

R

Raised and Rooted (Tyson Foods) 44
Rapid Trait Development System (RTDS™) 60
reductionism 6, 20, 36
Reeve, E. 84
remote sensing techniques 13
re-scripting
  concept of 7, 9, 29
  evidence, from UK 27–8
responsible innovation 1–2, 33
'right to repair' controversy 9, 30, 99n11
Roesch, F. 83
Rohlilk 103n3
Rose, D.C. 7, 27–8, 29
Rotz, S. 31
Russia–Ukraine war 1, 96

S

Safeway 73, 75
Sainsbury's 37, 104n17
Santo, R. 41
Schimmelpfennig, D. 25, 27
script, concept of 27
  see also re-scripting
seed–chemical complex 8, 9–10, 12, 18, 19
  and agro-ecosystem management 64
  and vertical integration 54–5
Sexton, A. 47
single cell protein (SCP) 37
Sinochina 98n13
smart farming 13, 26, 30
  see also precision agriculture
socio-technical transitions, and knowledge production 34
soil monitors 12, 25
Solum 98n10
Sonka, S. 13
Sorj, B. 5
soy leghemoglobin 40, 100n9

space 6, 35
Spackman, C. 50
spatial workflows, re-scripting of 28
Srnicek, N. 15
Starbucks 42
Starbucks China 45
Steinbrecher, R.A. 59
Stephens, N. 49, 50
Stoller, M. 84
Strider 18
Structural One Health 3
Substitutionism 6
Substitutionism 4.0 50–2
Summit Scientific Group 93
sustainability transition, possibility of 90–1
sustainable intensification, importance of 19–20
Sweet Earth Foods 44
Sykuta, M.E. 15
Syngenta Group 13, 16, 18, 19, 54, 57, 64, 68, 98n13
synthetic gene drive systems 62
Sysco 42

T

Takeaway 72, 81
TakeawayJustEat 76, 77, 80, 81, 103n9, 104n2 (Ch 7), 104n12 (Ch 6)
TALEN 60, 102n14
Target 42, 73
Taylorism, new 99n17
Technion-Israel Institute of Technology 39
techno-centric path dependence redux 93–5
Tencent 15
Tesco UK 87, 104n17
Tilney, M. 18
tissue engineering systems 38–9, 41, 49
Trump Administration 25, 55, 63, 68–9
Trussell Trust 84
Tyson Foods 4, 44, 45, 46

U

Uber 83
Uber Eats 76, 77, 79, 80, 81
Uber-isation 72
Unilever 44, 45
United Food and Commercial Workers (UFCW) union 75
United Kingdom
  dark stores in 72
  Food Standards Agency 84
  gig economy 82
  groceries and food delivery in 81–2, 104n2
  impact of COVID-19 pandemic in 84–5
  inflation rate in 96
  re-scripting technologies and farmers in 27–8
United Nations
  Committee on World Food Security (CFS) 91, 92, 104n5

Conference of the Parties (COP26) 11, 93, 94, 96
Food and Agriculture Organisation (FAO) 2–3, 92
Food Systems Summit (UNFSS) 11, 92–3, 94, 105n8, 105n9
United States
  adoption of precision agriculture in 25–6
  alternative proteins industry in 42–3, 45–6, 48
  Bayh-Dole Act (1980) 65
  biotechnology in 55, 57–8
  California Assembly Bill 5 (AB5) 83
  California's Proposition 22 82–4, 86, 104n13
  Crisis by Design report 5, 21–2
  dairy farms in 24
  dark stores in 72
  Department of Agriculture (USDA) 25, 61, 66, 67, 68–9
  Digital Millennium Copyright Act (DMCA) (1998) 31
  Environmental Protection Agency (EPA) 61, 63, 66, 69, 101n6, 102n17
  farm crisis 5
  farm policy 88–9
  gene drives in 63
  grocery chains in 73, 74
  impact of COVID-19 pandemic in 84, 85
  livestock industry in 23, 58
  meal delivery companies in 77
  Mid-West 2, 5, 25, 54
  National Academies of Sciences, Engineering and Medicine (NASEM) 63
  online grocery delivery in 74–5, 80
  regulation of agri-biotechnology 64–5, 66, 67, 68–9
  SECURE rule 68
  structural fault lines and change in 21–3
Upside Foods 41, 44
utility patents legislation 65

V

variable rate technology (VRT) 13, 16, 25
Veen, A. 83
Vegetarian Butcher 44

venture capitalism 8, 12, 18, 53, 98n12
  and alternative proteins 36–8, 42, 43, 47, 48, 52
  corporate examples of 100n13
  and platform capitalism 81
  promissory narratives for 9
  and supermarkets 90
vertical integration 17, 30
  in app-based food delivery 77, 78
  and seed-chemical complex 54–5
  see also bundling
Vivera 45

W

Waitrose 104n17
Walmart 73, 75, 86
Watkins, L. 15, 78, 80, 81
Weezy 74, 104n18
Weis, T. 5
Wellesley, L. 41
Wen, L. 85
White Castle 42
WhiteWave Foods 43
Whole Foods 42, 72, 73, 74
Whoosh 104n17
Wiens, K. 30
Wilkinson, J. 5
Winter, A. 26
Wiseman, L. 30
Wolf, S. 12, 89
World Economic Forum (WEF) 53, 92
World Food Programme (WFP) 104n16

X

Xingsheng Youxuan 103n3

Y

Yglesias, M. 75
yield monitoring 12, 13, 14, 28
Yum Brands 42

Z

ZFN 60, 102n14
zoonotic diseases 3, 61
  see also COVID-19 pandemic

www.ingramcontent.com/pod-product-compliance
Lightning Source LLC
Chambersburg PA
CBHW071714020426
42333CB00017B/2264